JCNS Workshop on Modern Trends in Production and Applications of Polarized 3He 2010

Journal of Physics: Conference Series Volume 294

Munich-Ismaning, Germany
11 – 13 July 2010

Editors:

Alexander Ioffe
Earl Babcock
Thomas Gutberlet

ISBN: 978-1-61839-003-5
ISSN: 1742-6588

Printed from e-media with permission by:

Curran Associates, Inc.
57 Morehouse Lane
Red Hook, NY 12571

Some format issues inherent in the e-media version may also appear in this print version.

Printed by Curran Associates, Inc. (2011)

For permission requests, please contact the Institute of Physics
at the address below.

Institute of Physics
Dirac House, Temple Back
Bristol BS1 6BE UK

Phone: 44 1 17 929 7481
Fax: 44 1 17 920 0979

techtracking@iop.org

Additional copies of this publication are available from:

Curran Associates, Inc.
57 Morehouse Lane
Red Hook, NY 12571 USA
Phone: 845-758-0400
Fax: 845-758-2634
Email: curran@proceedings.com
Web: www.proceedings.com

TABLE OF CONTENTS

INVITED CONTRIBUTIONS

CELLS AND CELL MATERIALS

DEVICES FOR PRODUCTION AND USE OF POLARIZED ^{3}HE

PRECISION ^{3}HE POLARIMETRY AND METHODS

POLARIZED ^{3}HE TARGETS AND APPLICATIONS

PREFACE

Polarized neutron scattering techniques are an indispensable and highly requested tool for studying magnetic phenomena in condensed matter. The different coherent and incoherent scattering of isotopes such as protons and deuterons also allows applications of polarized neutrons in soft matter and biological studies of molecular and macromolecular dynamics.

One method to polarize neutrons is to use polarized ^{3}He gas which absorbs, or filters, one spin state of the neutron beam as it passes through it. Only about ten years ago, early polarized neutron scattering experiments using such ^{3}He neutron spin filters (^{3}He NSF) were being conducted using starting ^{3}He polarizations of 55%. Currently there are two different commonly used methods to polarize high quantities of ^{3}He. These methods both collisionally transfer spin polarization to ground state ^{3}He nucleuses; one method uses optical pumping of an excited metastable state of ^{3}He atoms, and the other uses optical pumping of the ground state of an alkali-metal vapour. Within the last decade immense progress in both methods has resulted in ^{3}He polarizations of up to 80% being reported in atmosphere-pressure ^{3}He cells by the world's leading labs.

This progress in optical pumped ^{3}He promises to give rise to much more efficient and novel polarized neutron scattering experiments as and also impacts other areas of science. Polarized ^{3}He is additionally applied in research fields such as particle physics, fundamental studies and medicine. Thus not only the techniques and methods of polarization, but the research groups themselves exploring polarized ^{3}He, have a large breadth and diversity spanning different fields of science and locations in the world. Given this diversity, it is rare for this community to meet as a group at any one meeting or conference.

Because it is crucial to discuss new developments in ^{3}He polarization in a multi-disciplinary international setting, an international workshop on "Modern Trends in Production and Applications of Polarized ^{3}He" was held from 11–13 July 2010 in Ismaning, Germany. This workshop was organized by the Jülich Centre for Neutron Science (JCNS) as part of our ongoing series of topic-oriented JCNS workshops relating to specific activities of the JCNS. In particular this JCNS workshop discussed the techniques of polarization of ^{3}He, ^{3}He containers, magnetic holding systems, ^{3}He spin manipulations and applications in neutron scattering, medicine and physics.

The workshop was very well attended with intense discussions and inspiring presentations. In many ways the current workshop was a conceptual descendent of the HELION workshop in 2002 in Mainz, Germany. It is the hope of both us and many of the participants that such a workshop will be held again, perhaps 3 to 4 years in the future.

The current proceedings is a collection of papers written within the spirit of the workshop to provide a documentation of the current status and perspectives in the polarization of ^{3}He and its applications.

Alexander Ioffe
Earl Babcock
Thomas Gutberlet
Editors

JCNS Workshop on Modern Trends in Production and Applications of Polarized ^{3}He IOP Publishing
Journal of Physics: Conference Series **294** (2011) 012001 doi:10.1088/1742-6596/294/1/012001

Fundamentals of Spin-Exchange Optical Pumping

Thad G. Walker
Department of Physics, University of Wisconsin-Madison, Madison, WI USA 53706
E-mail: tgwalker@wisc.edu

Abstract. The standard model of Spin-Exchange Optical Pumping of ^{3}He has come under close scrutiny as researchers try to maximize the quantity and polarization of ^{3}He. This paper will review four important effects that have come to light: 1) pumping with frequency narrowed diode lasers significantly improves the polarization and production rate as compared to off-the-shelf broadband sources 2) Rb-K mixtures again improve the production rate due to reduced collisional loss in K-He collisions, 3) a unexplained ^{3}He wall relaxation mechanism that scales closely with alkali vapor pressure limits the attainable ^{3}He polarization to about 80%, 4) a small collisionally-induced breakdown in the selection rules for optical pumping explains why narrowband lasers are so advantageous, and may explain some observed limitations for Rb-K pumping as well.

1. Introduction

The Spin-Exchange Optical Pumping (SEOP) method for producing spin-polarized ^{3}He is in principle relatively simple and inexpensive to implement experimentally, yet still provides excellent performance. The basic processes governing SEOP were reviewed a number of years ago [1], and are summarized by the following equations:

$$\frac{dP_{\rm He}}{dt} = k_{\rm SE}[{\rm A}]\,(P_{\rm A} - P_{\rm He}) - \Gamma_{\rm He}P_{\rm He} \tag{1}$$

$$P_{\rm A} = \frac{R(z)}{R(z)+\Gamma_{\rm A}} \tag{2}$$

$$\Gamma_{\rm A} = (k_{\rm SE}+k_{\rm SR})\,[{\rm He}] + \Gamma_0 \tag{3}$$

$$\frac{dR(z)}{dz} = -[{\rm A}]\sigma_L\Gamma_{\rm A}P_{\rm A} \tag{4}$$

Equation 1 states that polarization is transferred back and forth by spin-exchange collisions between the alkali atoms of polarization $P_{\rm A}$ and the ^{3}He nuclei of polarization $P_{\rm He}$. The transfer rate, proportional to the alkali density [A], is typically 0.1/hr. The ^{3}He polarization is also lost to the walls of the container at a rate $\Gamma_{\rm He}$ which at room temperature is now routinely less than 0.005/hr [2]. Assuming no temperature dependence to the wall relaxation rate, Eq. 1 predicts that the steady-state polarization should be $P_{\rm He} > 0.95P_{\rm A}$.

Equation 2 states that the alkali polarization at position z in the cell is a competition between optical pumping (rate $R(z)$) and alkali spin-relaxation processes (rate $\Gamma_{\rm A}$). The polarization is high as long as $R(z) \gg \Gamma_{\rm A}$. The relaxation processes, Eq. 3, include spin-exchange ($k_{\rm SE}$) and spin-relaxation ($k_{\rm SR}$) collisions with ^{3}He, and other processes such as relaxation in alkali-alkali collisions and diffusion to the walls.

Equation 4 states that the pumping rate, which is proportional to the light intensity, decreases as the light propagates through the optically pumped vapor. As long as the alkali polarization is nearly 1, the attenuation is linear with distance. Absent the optical pumping to the dark state, the absorption length of $1/[\mathrm{A}]\sigma_L \approx 50$ μm for an atomic absorption cross section of $\sigma_L \approx 5 \times 10^{-13}$ cm^2 in a [He]=1 atm cell implies that the light would be completely absorbed less than a mm after entering the cell.

Now let us estimate the performance that should be attained. We assume $P = 10$ W of light entering a 10 cm diameter, 4.5 cm long cylinder with 1 atm of ^{3}He. The vapor is Rb, at a density of 4×10^{14} cm^{-3}. The pumping rate entering the cell of area A is $R(0) = \sigma_L P/(\hbar\omega A) = 2.6 \times 10^5$/s. The estimated spin-relaxation rate is $\Gamma_{\mathrm{A}} = 270$/s, mostly from Rb-Rb collisions [3]. At the exit of the cylinder, numerical solution of Eq, 4 finds $R(4.5) = 0.05R(0)$, with $P_{\mathrm{A}} = 0.978$. The cell-averaged polarization is in excess of 99%. The ^{3}He polarization should be in excess of 95%.

In practice, in order to successfully pump a cell of this size requires 50 W or more of frequency narrowed diode laser light [3], and the ^{3}He polarization is found to be at most in the high 70%s. When Ref. [1] was written, the polarizations obtained with high power diode arrays were typically in the mid-50% range or lower. In this paper I will summarize some of the effects that have been discovered and some of the experimental techniques that have been introduced in our laboratory at Wisconsin, working closely with our colleagues at NIST, to improve this situation.

2. Frequency-narrowed lasers

Wagshul and Chupp [4] first showed that it is possible to use high power semiconductor lasers for SEOP of ^{3}He. Previously, ion-laser pumped dye lasers had been used. Comparatively, semiconductor lasers offer greatly reduced price and greatly increased reliability, crucial for stable operation over the 10s of hours required to produced the polarized ^{3}He. The principal drawbacks were poor temporal and spatial mode structures of these lasers. In particular, the linewidths of these lasers tended to be several nm corresponding to 1 THz or greater in frequency units. When compared to the 20 GHz pressure-broadened linewidth at 1 atm ^{3}He pressure, this appeared to be a significant disadvantage. However, since SEOP cells are extremely optically thick at operating temperatures, light in the spectral wings of the laser is still absorbed, albeit at a smaller rate [5]. Even when high power arrays were used, it was found that polarizations were at most 60%.

In the late '90s, use of frequency narrowed semiconductor lasers began to be pursued. Levron *et al.* [6] used a custom laser for CsXe SEOP. In our laboratory, Ian Nelson and Bien Chann showed that off-the-shelf diode arrays [7] and diode-array bars [8] could be frequency narrowed, and with Tom Gentile's group at NIST showed that the polarizations dramatically improved, into the mid 70% range [9], as shown in Fig. 1.

The linewidth of these sources, typically 50-100 GHz is well-matched to the lineshape for optical pumping at 5-10 atm, but is still broader than would be best for neutron physics applications that prefer on the order of 1 atm. Thus the recent demonstration of Knize's group [10] of < 2 GHz linewidth for an external cavity laser with a high quality AR coated laser chip is intriguing for future improvements.

3. Hybrid Spin-Exchange Optical Pumping

In 1998, Baranga *et al.* [11] showed that the fundamental spin-exchange and spin-relaxation mechanisms for K-^{3}He are significantly favorable as compared to Rb-^{3}He. In practice, however, diode lasers for K are generally more difficult to obtain and are less developed than for Rb. For this reason, we demonstrated a hybrid optical pumping scheme in 2003 [12] that potentially allows for the best of both worlds: a high density K vapor is polarized by efficient spin-exchange collisions with a low density optically pumped Rb vapor.

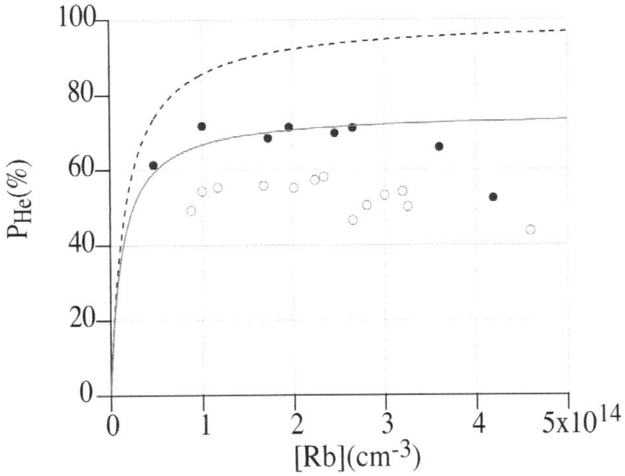

Figure 1. Comparison of ^{3}He polarizations obtained with broadband (42 W, open circles) and frequency narrowed light (14 W, closed circles) for ^{3}He SEOP. The solid line shows the calculated polarization including an unknown excess relaxation (see Sec. 3). The dotted line shows the predicted polarization assuming 100% Rb polarization and ^{3}He relaxation solely due to spin-exchange and room-temperature wall relaxation. From Ref. [9].

The advantages of hybrid spin-exchange can be understood with the following argument. The polarization achieved in the Rb is as always a balance between the optical pumping rate and the spin-relaxation rate. But in a hybrid cell each Rb atom has an effective spin-relaxation rate that is the sum of its direct relaxation rate and the relaxation of $\mathcal{D} = [\mathrm{K}]/[\mathrm{Rb}]$ K atoms. Thus the spin-polarization (Eq. 2) for hybrid spin-exchange becomes

$$P_A = \frac{R}{R + \Gamma_{\mathrm{Rb}} + \mathcal{D}\Gamma_{\mathrm{K}}} \tag{5}$$

The photon demand, number of photons scattered per second per alkali atom, is

$$\phi = \frac{R(1-P_{\mathrm{A}})}{1+\mathcal{D}} = \frac{\Gamma_{\mathrm{Rb}} + \mathcal{D}\Gamma_{\mathrm{K}}}{1+\mathcal{D}} P_{\mathrm{A}} \tag{6}$$

which is less than the value $\Gamma_{\mathrm{A}} P_{\mathrm{A}}$ for pure Rb pumping. With increasing K density, the additional relaxation increases more slowly than the atom density.

The increased capacity of hybrid spin-exchange can in principle be used to increase the spin-exchange rate (Fig. 2) or increase the polarized volume. However, it was discovered in the original hybrid experiments [12] that P_{He} begins to decrease for large $\mathcal{D}$. This was speculated to be a result of the K atoms having a non-negligible probability for absorbing the optical pumping light. This highly-off-resonant absorption is likely insensitive to the K spin-polarization, and therefore acts as an effective spin-relaxation mechanism. Experiments to quantify this possibility are currently being made by Brian Lancor.

A careful study of the gains from hybrid spin-exchange was led by Wangchun Chen and Tom Gentile [3]. Taking into account light propagation effects and tradeoffs between polarized volume and speed, it was found that the range $\mathcal{D} \sim 2-6$ generally gives the best performance.

4. He polarization limits

Models of the SEOP process, using known spin-exchange rates and measured room-temperature wall-relaxation rates, consistently predict $P_{\mathrm{He}} > 0.9$, whereas the highest values measured to

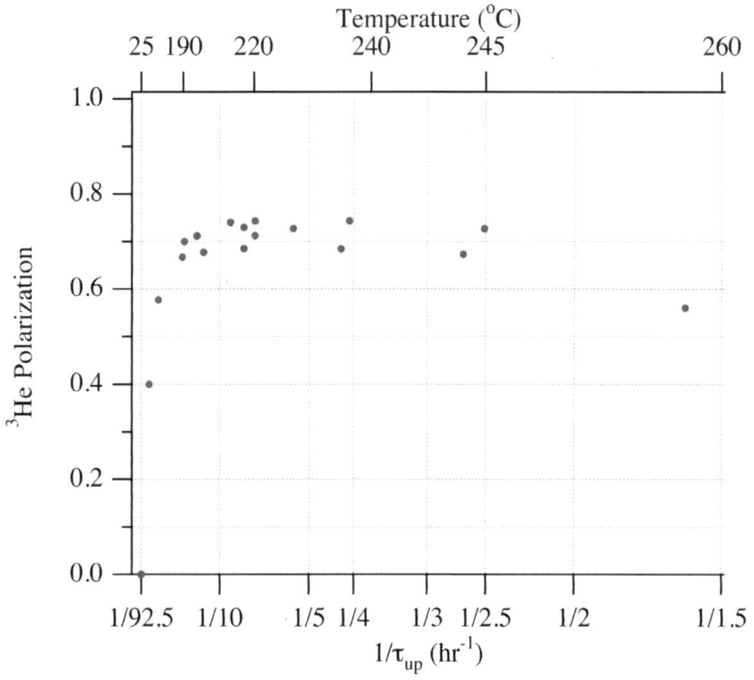

Figure 2. Fast spin-exchange optical pumping of a 25 cm^3, 3.3 amagat hybrid $\mathcal{D} = 1.5$ Rb/K cell using a frequency-narrowed diode laser. From Ref. [12].

date are about 0.8. From Eq. 1, the ^{3}He polarization should be

$$P_{\rm He} = \frac{k_{\rm SE}[{\rm A}]P_{\rm A}}{k_{\rm SE}[{\rm A}] + \Gamma_{\rm He}} \tag{7}$$

At high alkali densities, this should approach $P_{\rm A}$ unless $\Gamma_{\rm He}$ also increases at high [A]. The first indication that $\Gamma_{\rm He}$ has a strong temperature dependence was uncovered in Bien Chann's measurements of $k_{\rm SE}$ for Rb [13], where the values of $k_{\rm SE}$ deduced from wall-independent methods were found to be consistently smaller than indicated from the traditional method of studying the temperature dependence of spin-exchange time constants.

A systematic study of this "X-factor" effect was undertaken by the NIST and Wisconsin groups [14]. The wall-relaxation rate of ^{3}He was indeed found to vary as

$$\Gamma_{\rm He} = \Gamma_0 + k_{\rm SE}[{\rm A}]X \tag{8}$$

where X is a parameter that was found to be 0.2 or greater, but to fluctuate wildly from cell to cell. The fluctuations scale with surface-to-volume ratio S/V, as seen in Fig. 3:

$$X = X_0 + \chi\frac{S}{V} \tag{9}$$

where χ is a poorly controlled parameter, varying randomly from cell to cell. Large cells have smaller fluctuations than small ones. At high alkali density, the polarization limit becomes

$$P_{\rm He} < \frac{1}{1+X} \tag{10}$$

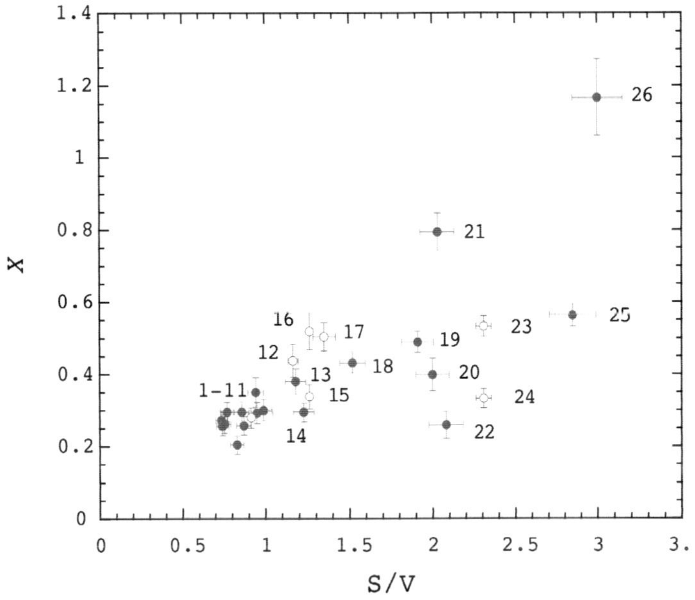

Figure 3. Measured X-factors for a variety of SEOP cells, done at NIST. From Ref. [14], where the characteristics of the various cells, identified here by number, are listed.

thus explaining why the X-factor causes the He polarization to reach a hard limit at high alkali densities.

A histogram of the data from Ref. [14], reparameterized in terms of χ, is shown in Fig. 4. The distribution is approximately Gaussian, with a mean of $\bar{\chi} = 0.31$ cm and a standard deviation of $\sigma_\chi = 0.08$ cm. This seems to support the hypothesis that the X-factor is linearly correlated with surface-to-volume ratio.

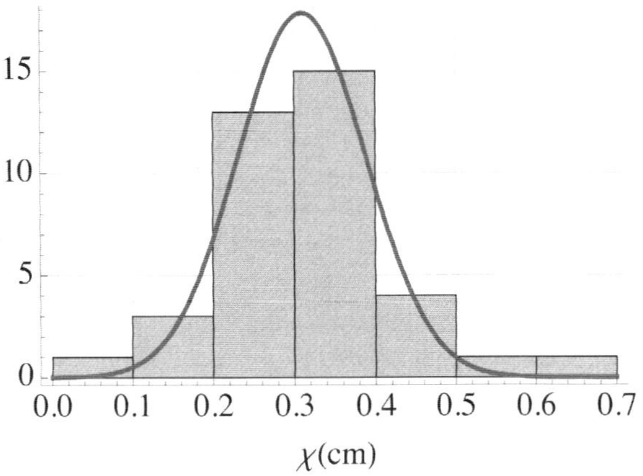

Figure 4. Distribution of χ-values, where $X = \chi S/V$, for the 38 cell sample reported in Ref. [14]. The solid line is a best-fit Gaussian distribution with a mean of 0.31 cm and a standard deviation of 0.08 cm.

The lack of cells with $X < 0.2$ brings up the question of whether or not there might be a fundamental process contributing to X. The only known such process is anisotropic spin-exchange [15]. Experimentally demonstrating anisotropic spin-exchange is difficult, as the effects are difficult to disentangle from wall effects when making measurements on ^{3}He relaxation, or appear simply as increased spin-relaxation rates when measuring alkali relaxation. A recent experiment [16] introduced a method that in principle can unambiguously isolate the effect, but was not sensitive enough to make a measurement of the effect: $X_0 = 0.10 \pm 0.13$.

5. Light-Induced Relaxation

In the first hybrid experiment [12], we observed that for broad-band pumping it was impossible to fully polarize the alkali atoms, even at low alkali density and very high laser intensities. This brought up the possibility that in the presence of high pressure He fully polarized alkali atoms might still absorb the circularly polarized pumping light, contrary to the selection rules for free atoms. Due to rapid excited-state spin-relaxation, such a process would act to *de-pump* the atoms and lead to a maximum alkali polarization P_∞ at high pumping rates. Even if P_∞ is very close to 1, the extreme optical thickness in SEOP cells can make this process very damaging. This is because the light-induced relaxation with cross-section σ_∞ adds to Eq. 4 a standard Beer's law type term:

$$\frac{dR(z)}{dz} = -[\mathrm{A}]\sigma_L\Gamma_\mathrm{A}P_\mathrm{A} - [\mathrm{A}]\sigma_\infty R(z) \tag{11}$$

High optical depths and alkali polarizations require $R \gg \Gamma_A$, causing $\sigma_\infty R$ to be comparable to or even greater than Γ_A.

Initial measurements of this effect were made for off-resonant pumping light by Earl Babcock in his thesis [17], with a careful, complementary on- and off-resonant study being done by Brian Lancor [18, 19]. Such measurements are equivalent to studying the spin-dependence of pressure-broadening, and much of the molecular physics had been studied before in the context of fine-structure changing collisions [20]. Nevertheless, to our knowledge the effect of pressure-broadening collisions on the angular momentum selection rules had never been observed experimentally before.

Figure 5 shows how the light-induced absorption affects light propagation. The power and polarization through the cell are shown. As the light traverses the cell, the off-resonant pumping increasingly favors σ_∞ as compared to σ_L. The Rb polarization drops much more quickly with the dichroism effect than would be expected just from normal off-resonant pumping. We believe that this explains for the first time why such large polarization increases were observed when implementing narrowband pumping; the narrowband pumping light not only increased the pumping rate but also decreased the relative importance of σ_∞ as compared σ_L.

6. Conclusions

While considerable progress has been made in improving both the performance and understanding of SEOP over the past decade, important problems remain. Gauging from the response of the attendees at the JCNS Workshop, a solution to the X-factor problem would be most desirable. It would be nice to know if the observed X_0 arises from anisotropic spin-exchange. On this problem we are encouraged to note that the ITAMP group has become quite interested in *ab-initio* calculations of spin-dependent interactions of van-der-Waals molecules [21] and they may soon have a theoretical preduction for X_0. In my opinion it is unlikely that the result will be larger that 0.1. We would also like to improve on the experiment of Ref. [16] to at least get a measurement of X_0 for KHe. The much larger spin-relaxation rates for Rb-He make such an approach extremely difficult for RbHe.

Concerning the surface contribution to X, we note that the materials that have been studied to date have all been glasses (really, glasses coated with alkali films) such as pyrex, quartz, GE180, 1722, and sol-gels. Of course it is necessary to have glass surfaces to allow the light into

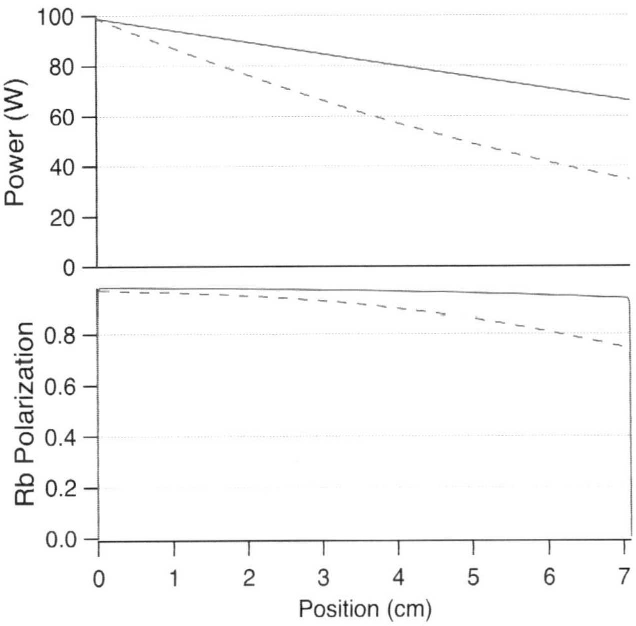

Figure 5. Optical pumping by a broad pump laser without (red, solid) and with (green, dashed) dark-state absorption. The upper graph shows that the dark state absorption produces a faster attenuation of the light. The lower graph shows how the dark state absorption reduces the Rb polarization at the cell entrance, with further reductions as the spectral profile of the light becomes increasingly off-resonant deeper inside the cell. From Ref. [18].

the cell. An example of an interesting experiment to try would be to coat the cell interior with a sub-skin-depth layer of gold or copper to see what differences that might make. Unfortunately, there is no model for the surface physics that gives rise to the X-factor, so it is difficult to know what particular metals might be promising.

We are currently engaged in a new assessment of the "photon budget" for SEOP. Given what we now know about light-induced relaxation and hybrid spin-exchange, along with other topics such as excited-state relaxation [22], radiation trapping [22], and cell heating [23], we hope that we are close to being able to model the SEOP process sufficiently accurately to account for the amount of light required to polarize ^{3}He.

Acknowledgments
I would like to thank I. Ioffe, E. Babcock, and T. Gutberlet for organizing the excellent JCNS Workshop on "Modern Trends in Production and Applications of Polarized 3-He". The work summarized here was performed by a long series of excellent students and collaborators. At Wisconsin, S. Kadlecek, I. Nelson, B. Chann, E. Babcock, and B. Lancor have all made extremely important contributions to SEOP of ^{3}He. We have enjoyed a long and fruitful collaboration with T. Gentile and his group at NIST. Our early work on these problems was supported by the National Science Foundation, with the recent work supported by the U.S. Department of Energy, Office of Basic Energy Sciences, Division of Materials Sciences and Engineering under award DE-FG02-03ER46093.

References

[1] Walker T G and Happer W 1997 *Rev. Mod. Phys.* **69** 629

[2] Rich D R, Gentile T R, Smith T B, Thompson A K and Jones G L 2002 *Appl. Phys. Lett.* **80** 2210
[3] Chen W C, Gentile T R, Walker T G and Babcock E 2007 *Phys. Rev. A* **75** 013416
[4] Wagshul M E and Chupp T E 1989 *Phys. Rev. A* **40** 4447–4454
[5] Driehuys B, Cates G, Miron E, Sauer K, Walter D and Happer W 1996 *Appl. Phys. Lett.* **69** 1668
[6] Levron D, Walter D K, Appelt S, Fitzgerald R J, Kahn D, Korbly S E, Sauer K L, Happer W, Earles T L, Mawst L J, Botez D, Harvey M, DiMarco L, Connolly J C, Möller H E, Chen X J, Cofer G P and Johnson G A 1998 *Applied Physics Letters* **73** 2666–2668
[7] Nelson I A, Chann B and Walker T G 2000 *Appl. Phys. Lett.* **76** 1356
[8] Chann B, Nelson I and Walker T G 2000 *Opt. Lett.* **25** 1352
[9] Chann B, Babcock E, Chen W, Smith T, Thompson A, Anderson L W, Walker T G and Gentile T R 2003 *J. Appl. Phys.* **94** 6908
[10] Sell J F, Miller W, Wright D, Zhdanov B V and Knize R J 2009 *Applied Physics Letters* **94** 051115
[11] Ben-Amar Baranga A, Appelt S, Romalis M V, Erickson C J, Young A R, Cates G D and Happer W 1998 *Phys. Rev. Lett.* **80** 2801
[12] Babcock E, Nelson I, Kadlecek S, Driehuys B, Anderson L W, Hersman F W and Walker T G 2003 *Phys. Rev. Lett.* **91** 123003
[13] Chann B, Babcock E, Anderson L W and Walker T G 2002 *Phys. Rev. A* **66** 032703
[14] Babcock E, Chann B, Walker T G, Chen W C and Gentile T R 2006 *Phys. Rev. Lett.* **96** 083003
[15] Walter D K, Happer W and Walker T G 1998 *Phys. Rev. A* **58** 3642
[16] Walker T G, Nelson I A and Kadlecek S 2010 *Phys. Rev. A* **81** 032709
[17] Babcock E 2005 *Spin-Exchange Optical Pumping with Alkali-Metal Vapors* Ph.D. thesis University of Wisconsin-Madison
[18] Lancor B, Babcock E, Wyllie R and Walker T G 2010 *Phys. Rev. Lett.* **105** 083003
[19] Lancor B, Babcock E, Wyllie R and Walker T G 2010 *Phys. Rev. A* **82** 043435
[20] Allard N and Kielkopf J 1982 *Rev. Mod. Phys.* **54** 1103–1182
[21] Tscherbul T V, Zhang P, Sadeghpour H R and Dalgarno A 2009 *Phys. Rev. A* **79** 062707
[22] Lancor B and Walker T G 2010 *Phys. Rev. A* **82** 043417
[23] Walter D K, Griffith W M and Happer W 2001 *Phys. Rev. Lett.* **86** 3264

Journal of Physics: Conference Series **294** (2011) 012002 doi:10.1088/1742-6596/294/1/012002

Fundamentals of metastability exchange optical pumping in helium

M Batz[1,2], P-J Nacher[1*] and G Tastevin[1]

[1] Laboratoire Kastler Brossel, ENS; CNRS; UPMC; 24 rue Lhomond, 75005 Paris, France.
[2] Johannes Gutenberg University, Staudingerweg 7, 55128 Mainz, Germany.

E-mail: `nacher@lkb.ens.fr`

Abstract. Advances in metastability exchange optical pumping (MEOP) at high laser powers, but also at high gas pressures and high magnetic field strengths, has provided strong motivation for revisiting the understanding of the limitations of this powerful technique. A comprehensive model has been developed for improved description of the combined effects of OP, ME, and relaxation, and of detailed MEOP features observed over the broad range of operating conditions. A brief description is provided, with illustrative comparisons of computed and experimental results. This improved tool is used to explain the excellent photon efficiency of OP obtained at all field strengths. It is combined with an angular momentum budget approach to quantitatively investigate the newly discovered strong OP-enhanced polarisation losses that currently limits MEOP performance.

1. Introduction

Metastability exchange optical pumping (MEOP), discovered in 1963 [1], exclusively applies to ^{3}He as demonstrated by various attempts for other gases [2, 3, 4], while spin-exchange OP (that involves an alkali vapour [5, 6, 7]) can be used to efficiently polarise ^{129}Xe as well. MEOP is actually performed between two excited states of helium (Fig. 1). The 2^3S metastable state, that plays the role of a ground state for the OP process, is populated by electron collisions in a plasma discharge sustained in a gas at moderate pressure (usually up to a few millibars). For the ^{3}He isotope in this excited metastable state, an efficient coupling between the nucleus and the electrons (the hyperfine interaction) results in a strong entanglement of electronic and nuclear spin states. Therefore the OP-enforced optical orientation of the electronic angular momentum simultaneously induces nuclear orientation as well. This nuclear orientation is transferred through metastability exchange collisions to the much more numerous atoms having remained in the true (1^1S$_0$) ground state.

When laser light with adequate spectral characteristics at 1083 nm is used, MEOP provides very high nuclear polarisation (>0.7) with good photon efficiency (~1 polarised nucleus per absorbed photon) [8]. Given the development of suitable high power fibre lasers [9, 10, 11], the only drawback of this method is the limited range of operating pressures (of order 0.5-5 mbar [12]) for which a suitable plasma can excite metastable atoms. When a higher final pressure is needed, non-relaxing compression of the gas becomes necessary and this introduces demanding requirements on the gas handling and compressing devices. However it has recently been shown that the range of operating pressures can be extended to several tens or hundreds of mbar by performing MEOP in a high magnetic field, such as the 1.5 T field of a magnetic resonance

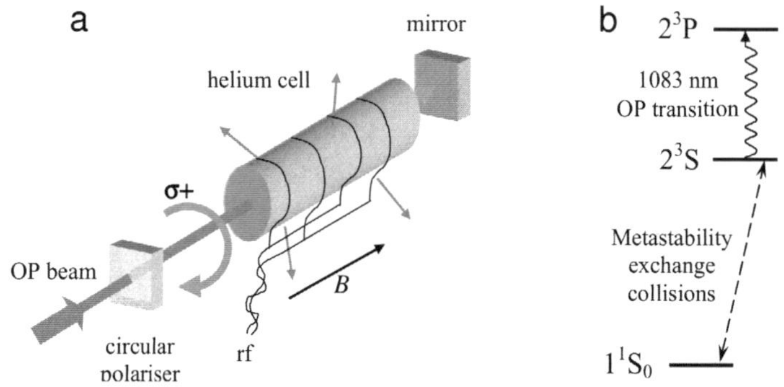

Figure 1. a: Schematic view of a basic MEOP set-up. In a ^{3}He gas cell, a weak rf discharge promotes a small fraction of the atoms into the excited metastable state 2^3S, where resonant absorption of 1083 nm light occurs. This OP light is circularly polarised and propagates along the direction of the applied magnetic field B. For enhanced absorption, the light is usually back-reflected for a second pass through the cell. Nuclear polarisation is transferred to atoms in the ground state by metastability exchange collisions. **b:** The atomic levels of He and the main physical processes involved in MEOP are shown. The 2^3S and 2^3P states are composed of several levels and the 1083 nm line thus has a complex structure (see Fig. 2).

imaging (MRI) system [13, 14], and up to 4.7 T [15]. Operating at higher pressures would make subsequent compression significantly easier to perform.

The first theoretical articles on MEOP were written in the 1970s [1, 16, 17, 18, 19]. At that time, OP on the 2^3S-2^3P transition was performed using the weak light from a helium lamp and rather low polarisations (of the order of 0.1) were obtained. Simple linearised models were sufficient to account for such pumping experiments. When laser sources at 1083 nm were developed in the 1980s, much higher nuclear polarisations were obtained. A detailed model for MEOP without restriction on the pump light intensity nor on the nuclear polarisation was then proposed [8]. It is based on rate equations for the populations of the 2^3S, 2^3P, and ground states, that accurately take into account the effects of metastability exchange collisions and of absorption and emission of light at 1083 nm. Correlations between atomic orientations and velocities, created for instance by a spectrally narrow laser in a room temperature gas, are also considered in this model, but only in a crude phenomenological way. This model only describes low field MEOP and it has been established and used for parameters corresponding to low pressure situations (in the mbar range) where MEOP was known to have the highest efficiency. This model has been also used to discuss OP experiments with a broadband laser [20, 21]. Extensions to specific situations have been proposed, but not fully developed, for OP in ^{3}He-^{4}He isotopic mixtures[1] [23] and for OP in high magnetic fields [12, 13, 24]. Recently, we have developed a more comprehensive model of MEOP. It is based on similar rate equations for the populations of all Zeeman sublevels but it is suitable for arbitrary magnetic field and for pure ^{3}He gas as well as for isotopic gas mixtures. It provides a unified frame that is also more adequate for description of MEOP with modern, broadband fibre lasers.

The main features of this improved model are described in section 2, where it is simply used as a tool to illustrate the combined effects of ME, OP, and relaxation for conditions typically met in MEOP-based ^{3}He gas polarisers. Section 3 is devoted to a more global discussion of the net growth rate of ^{3}He nuclear polarisation in the ground state, through an angular momentum

[1] OP performed on the 2^3S state of ^{4}He atoms in an isotopic mixture is an efficient scheme providing high ^{3}He nuclear polarisations [10, 22]. It can be advantageoulsly used when the presence of ^{4}He is desired or is not a nuisance, e.g., for low temperature studies of polarised mixtures or for neutron spin filters applications.

budget that combines the photon efficiency of the 2^3S-2^3P transition and a total polarisation loss that only involves 2^3S and ground state relaxation rates. Section 4 provides an example of experimental implementation and of analysis of polarisation dynamics for systematic MEOP study at low field. Comparison of measured and computed polarisations provides a clear evidence of a strong laser-induced relaxation rate that scales linearly with the absorbed light power.

2. Main features of the comprehensive MEOP model

The key processes in MEOP are the net transfer of angular momentum from absorbed light to spin degrees of freedom of He atoms (in the 2^3S and 2^3P states) and the transfer of nuclear angular momentum between ground state ^{3}He atoms and atoms in the 2^3S state by metastability exchange collisions. A detailed balance with the angular momentum losses resulting from various relaxation processes is used to derive rate equations for the relevant populations: $(1 \pm M)/2$ for the two Zeeman sublevels (m_I=$\pm 1/2$) of the ground state of ^{3}He atoms, characterised by its nuclear polarisation M, the populations of the 2^3S metastable state and those of the 2^3P state.

2.1. Level structure of the 2^3S and 2^3P states

We use the notations of reference [24] where the structure and energies of the sublevels of the 2^3S and 2^3P states are derived for both isotopes in arbitrary magnetic field. For simplicity we only discuss two experimentally relevant limiting cases: low and high magnetic fields.

2.1.1. Low field - The 2^3S state of ^{4}He has three magnetic sublevels (m_S=-1, 0, and 1), linearly split at all values of the applied magnetic field B by the Zeeman energy. They are named Y_1 to Y_3 and their populations y_1 to y_3 (for all sets of populations, indices increase with increasing energies). The 2^3P state of ^{4}He has three fine-structure levels with J=0, 1 and 2 (see Fig 2a), hence nine Zeeman sublevels. There are twice as many Zeeman sublevels for ^{3}He due to its two nuclear spin states: six (A_1 to A_6, populations a_1 to a_6) in the 2^3S state and eighteen (B_1 to B_{18}, populations[2] b_1 to b_{18}) in the 2^3P state that has five fine- and hyperfine-structure levels (see Fig 2b). In low magnetic field, the F=3/2 and F=1/2 hyperfine levels of the 2^3S state of ^{3}He are well resolved and split by 6.74 GHz (Fig 2b). The magnetic sublevels can be written using the decoupled basis states $|m_S, m_I\rangle$. A_1=$|-1,-\rangle$ and A_4=$|1,+\rangle$ are pure states of maximum $|m_F|$=3/2 but the other states involve large mixing parameters θ_- and θ_+ [24]:

$$\begin{aligned}
A_2 &= \cos\theta_- \,|-1,+\rangle + \sin\theta_- \,|0,-\rangle \\
A_3 &= \cos\theta_+ \,|0,+\rangle + \sin\theta_+ \,|1,-\rangle \\
A_5 &= \cos\theta_- \,|0,-\rangle - \sin\theta_- \,|-1,+\rangle \\
A_6 &= \cos\theta_+ \,|1,-\rangle - \sin\theta_+ \,|0,+\rangle\,. \qquad (1)
\end{aligned}$$

Strong and maximal mixing of electronic and nuclear angular momenta occurs at B=0, with $\sin^2\theta_-$=2/3 and $\sin^2\theta_+$=1/3. The mixing parameters and the sublevel energies, as well as those of the 2^3P_0 sublevels addressed by the C_8 and C_9 transitions used for ^{3}He MEOP (see Fig 2b), linearly depend on B for low fields. The corresponding Zeeman line shifts do not exceed 30 MHz/mT and the relative changes in transition intensities do not exceed 0.6 %/mT. Given the Doppler width of the optical transitions (of order 2 GHz fwhm at room temperature for ^{3}He), the OP rate equations are almost field-independent up to several mT.

[2] The local number density of 2^3S atoms in sublevel A_i is $n_m a_i$ (n_m is the number density of atoms in the metastable state and the a_is are populations such that $\sum_{i=1}^{6} a_i$=1). For convenience, the b_js are defined so that the number density of atoms in sublevel B_j of the 2^3P state is $n_m b_j$. Hence the b_js are not true populations, and $\sum_{k=1}^{18} b_k$<1 depends on the OP light intensity.

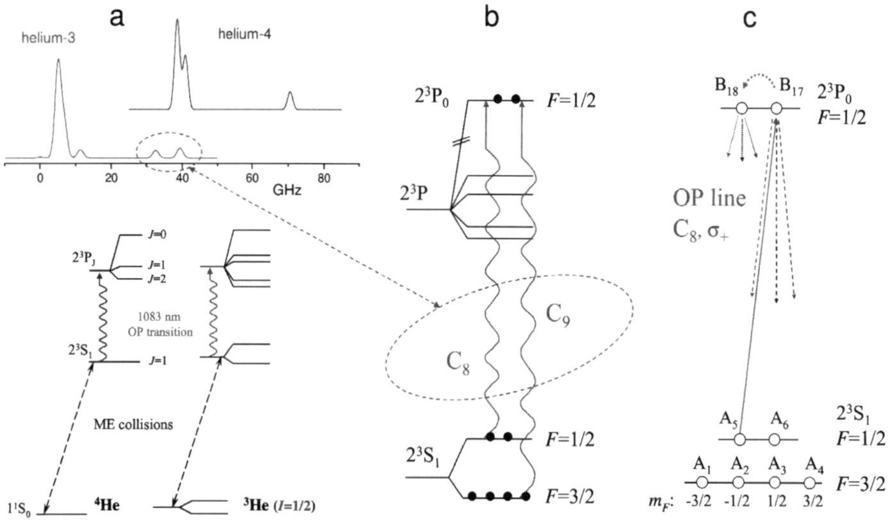

Figure 2. a: Fine- and hyperfine-structures of the atomic states of He involved in the MEOP process, for the ^{4}He (left) and ^{3}He (right) isotopes, in low magnetic field (below a few 10 mT). The resulting spectra of the 1083 nm transition (upper graphs, computed with 300 K Doppler widths and no collisional broadening) extend over ~70 GHz when the isotope shift is considered. Optical transition frequencies are referenced to that of the C_1 line at zero field [24]. **b:** C_8 and C_9 lines, which are the most efficient in usual OP conditions, connect the 2^3S_1, F=1/2 and F=3/2 levels of ^{3}He, respectively, to the 2^3P_0 level. **c:** Example of elementary processes considered in the angular momentum budget for C_8 OP (see text).

2.1.2. High field - When the Zeeman energy exceeds the fine- and hyperfine-structure energy scales the angular momentum structures of 2^3S and 2^3P levels and the 1083 nm transition are deeply modified. Figure 3 displays the energy of all Zeeman sublevels for B=1.5 T, a field strength commonly met in MRI systems and thus of practical importance for applications to human lung MRI using polarised ^{3}He gas. At this field strength, only a weak state mixing remains in the 2^3S state (Eqs. 1 with $\sin\theta_+$=0.07128 and $\sin\theta_-$=0.07697). The six Zeeman sublevels of the ^{3}He isotope levels are organised in three pairs of states (Fig. 3, left bottom graph). In each pair the level energy is mostly determined by the common dominant value of m_S (compare with Fig. 3, right bottom graph) while the nuclear spin projections are almost antiparallel. Similarly, hyperfine coupling only weakly mixes levels of different m_I values in the 2^3P state (Fig. 3, upper left graph). As a result, high-field spectra for a given light polarisation are thus composed of six main components which appear in two groups: a pair and a quartet, each group being unresolved at room temperature. These features clearly appear on the ^{3}He absorption spectra displayed in Fig. 4, that are computed at 1.5 T assuming room temperature Doppler widths and no collision broadening. The strong lines in the ^{3}He spectra are labelled $f_n^{\pm}$, where n=2 or 4 refers to the number of unresolved components and $\pm$ to the sign of the circular light polarisation. Weaker doublets of resolved single-component lines of interest for optical detection purpose are highlighted (two options exist for σ_+ polarisation). They separately address sublevels A_1 and A_2 (resp. A_5 and A_6) for σ_+ (resp. σ_-) polarised light.

2.2. Metastability exchange and spin-temperature distribution

An important feature of metastability exchange collisions in helium is that negligible depolarisation occurs during the collisions, due to the fact that all involved angular momenta

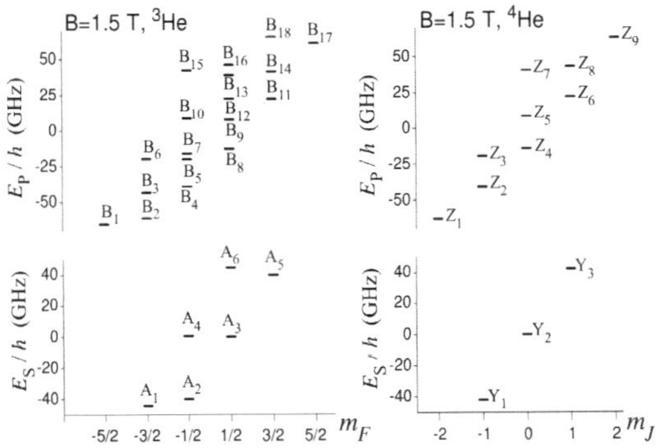

Figure 3. Energies (E_S/h and E_P/h, in frequency units) and angular momentum projections (m_F or m_J) of the sublevels of the 2^3S and 2^3P states of ^{3}He (left) and ^{4}He (right) at 1.5 Tesla. Energy references and state labelling are identical to those of [24].

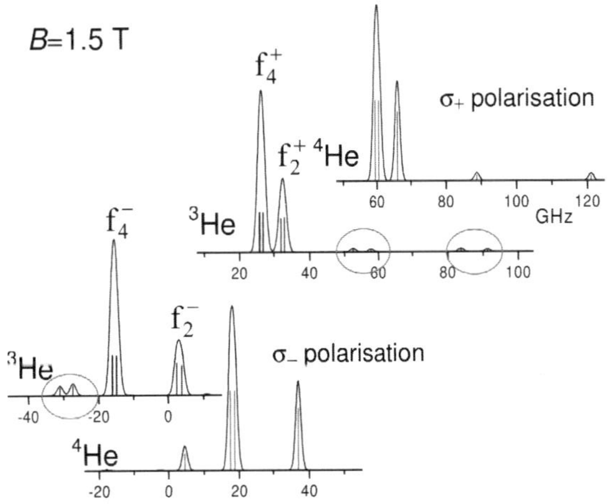

Figure 4. 1083 nm absorption spectra of ^{3}He and ^{4}He at 1.5 T for circularly polarised light. Optical transition frequencies are referenced to that of the C_1 line in zero field [24]. Pump and probe lines relevant for high field MEOP are described in text.

are spins [25]. Introducing density operators ρ_g and ρ_m to characterise the spin states of colliding ground state and metastable state atoms, respectively, a ME collision where excitation is exchanged between colliding atoms can be described as:

$$\begin{array}{ccccccc} \mathrm{He} & + & \mathrm{He}^* & \rightarrow & \mathrm{He}^* & + & \mathrm{He} \\ \rho_{\mathrm{g}} & & \rho_{\mathrm{m}} & & \rho'_{\mathrm{g}} & & \rho'_{\mathrm{m}}. \end{array} \tag{2}$$

All relevant collisions between ^{3}He and ^{4}He atoms occuring in isotopic mixtures are discussed in reference [24]. In pure ^{3}He, one simply has just before a collision:

$$\rho_{\mathrm{g}} = \frac{1+M}{2}\left|+\right\rangle\left\langle+\right| + \frac{1-M}{2}\left|-\right\rangle\left\langle-\right| \text{ and } \rho_{\mathrm{m}} = \sum\nolimits_1^6 a_i\left|\mathrm{A}_i\right\rangle\left\langle\mathrm{A}_i\right|, \tag{3}$$

and just after ME collision:

$$\rho'_{\mathrm{g}} = \mathrm{Tr}_e\,\rho_{\mathrm{m}} \text{ and } \rho'_{\mathrm{m}} = \rho_g \otimes \mathrm{Tr}_n\,\rho_{\mathrm{m}} \tag{4}$$

where Tr_e, Tr_n are trace operators over the electronic and nuclear variables respectively [19]. Indeed, ρ'_{m} contains off-diagonal terms and has to be projected on the eigenstates A_i to derive rate equations for the populations. These equations explicitly depend linearly on M due to the linear dependence of ρ_{g} on M in Eq. 4. For pure ^{3}He they reduce to:

$$\left.\frac{da_i}{dt}\right|_{\mathrm{ME}} = \gamma_{\mathrm{e}}\left(-a_i + \sum_{k=1}^{6}(E^3_{ik} + MF^3_{ik})a_k\right). \tag{5}$$

The ME collision rate γ_{e} is proportional to gas pressure (3.75×10^6 s^{-1}/mbar in pure ^{3}He [26]). The B-dependent matrices E^3 and F^3 involve the mixing parameters θ_+ and θ_- of Eqs. 1 [24]. Given the decreasing values of these parameters at high B, several elements in the matrices decrease so as to only leave weaker couplings of order $\theta^2_\pm$. However, the ME collision rate is increased at high pressures and this advantageously compensates for the weaker couplings in the ME rate equations.

It can be shown that the populations of the sublevels that are coupled by ME collisions follow a Boltzmann-like distribution of angular momenta in all situations where OP and relaxation processes have negligible effect on populations. This is valid at all field strengths, and for pure ^{3}He as well as for isotopic mixtures [24]. Since number densities n_m of atoms in the metastable state are much lower than the ground state number density N_g (typically, $n_\mathrm{m}/N_\mathrm{g} \sim 10^{-6}$), the spin temperature (ST) of this distribution is controlled by the nuclear polarisation M. Noting e^β the ratio $(1+M)/(1-M)$ of the populations in the ground state ($1/\beta$ is the ST), the populations a_i^ST and y_i^ST are given by [24]:

$$a_i^\mathrm{ST} = e^{\beta m_\mathrm{F}}/(e^{3\beta/2} + 2e^{\beta/2} + 2e^{-\beta/2} + e^{-3\beta/2}) \quad \text{for } ^3\text{He}, \tag{6}$$

$$y_i^\mathrm{ST} = e^{\beta m_\mathrm{S}}/(e^{\beta} + 1 + e^{-\beta}) \quad \text{for } ^4\text{He}, \tag{7}$$

where m_F and m_S are the relevant angular momentum projections. The ST populations can alternatively be expressed as functions of M. The following formulas are valid up to B=0.1619 T:

$$a_1^\mathrm{ST} = \frac{(1-M)^3}{2M^2+6},\ a_4^\mathrm{ST} = \frac{(1+M)^3}{2M^2+6} \quad \text{for } ^3\text{He},\ m_F = \mp 3/2, \tag{8}$$

$$a_2^\mathrm{ST} = a_5^\mathrm{ST} = \frac{(1-M)^2(1+M)}{2M^2+6},\ a_3^\mathrm{ST} = a_6^\mathrm{ST} = \frac{(1-M)(1+M)^2}{2M^2+6} \quad \text{for } ^3\text{He},\ m_F = \mp 1/2, \tag{9}$$

$$y_1^\mathrm{ST} = \frac{(1-M)^2}{M^2+3},\ y_2^\mathrm{ST} = \frac{1-M^2}{M^2+3},\ \text{and}\ y_3^\mathrm{ST} = \frac{(1+M)^2}{M^2+3} \quad \text{for } ^4\text{He},\ m_J = -1,\ 0,\ \text{and } 1. \tag{10}$$

Level crossings (A_4 and A_5 at 0.1619 T, A_3 and A_4 at 4.716 T [24]) simply result in the corresponding permutation of indices in Eqs. 8 and 9 at high field strengths.

The spin-temperature distribution of populations at given ground state polarisation is affected by two competing processes: relaxation in the 2^3S state, that tends to equalise all populations, and OP, that tends to deplete the sublevels addressed by the optical transitions. They are now briefly discussed for MEOP in pure ^{3}He gas.

2.3. *Relaxation processes*

Relaxation phenomenologically describes the effect of additional processes that couple populations of different sublevels. They involve very different time scales in the two excited states:

In the 2^3S state, collisions do not affect the orbital angular momentum L=0 and only couple populations through spin exchange (i.e., through ME that conserves angular momentum and gives rise to the contribution described by Eq. 5). Relaxation is thus usually rather slow, associated for instance with de-excitation of the atom and re-excitation in the plasma. The effect of this process on ^{3}He populations is taken into account in the MEOP model through:

$$\left.\frac{da_i}{dt}\right|_\mathrm{r} = \gamma_\mathrm{r}^\mathrm{S}\left(\frac{1}{6} - a_i\right). \tag{11}$$

Depending on pressure, cell dimensions, gas purity, etc., the relaxation rate $\gamma_\mathrm{r}^\mathrm{S}$ is typically of order 10^3 s^{-1} in the 2^3S state. A lower bound is set by de-excitation on the cell wall, which involves diffusion of metastable atoms in the gas (with a diffusion coefficient given in [27]).

In contrast, in the 2^3P state, J-changing collisions occur at a much higher rate, proportional to pressure and of order a few 10^7s^{-1}/mbar [28, 29]. This can result in significant population transfers during the 2^3P state radiative lifetime. They are phenomenologically described in the MEOP model by a single relaxation rate γ_r^P and by rate equations:[3]

$$\left.\frac{db_j}{dt}\right|_r = \gamma_r^P \left(\frac{1}{18}\sum_{k=1}^{18} b_k - b_j\right). \tag{12}$$

The 2^3P relaxation rate γ_r^P is expected to also scale with pressure p, and the value γ_r^P/p=0.32×10^7 s^{-1}/mbar has been recently inferred from OP efficiency measurements [30].

Equations similar to Eqs. 11 and 12 are given in [24] for the populations of ^{4}He atoms involved in MEOP of isotopic mixtures.

In high magnetic field, due to the deep change in the structure of the 2^3P state sublevels, this description using a single relaxation rate may be questioned. However, for MEOP at high enough pressure, fast redistribution among sublevels can be assumed to fully equalise the populations and a simplified model can be used (see [31] and Sec. 2.4).

2.4. OP processes - A two-class MEOP model

OP promotes atoms from selected Zeeman sublevels of the 2^3S state to sublevels of the 2^3P state. For instance, for the simple low-field case illustrated in Fig 2c, the C_8 line with right-handed circular (σ_+) polarisation excites atoms from the A_5 sublevel (m_F=-1/2) to the B_{17} sublevel (m_F=1/2). Radiative decay brings atoms back to the 2^3S state with well-defined branching ratios from the B_{17} sublevel (both by spontaneous and by stimulated emission, dashed lines in Fig. 2c) and from any other sublevel that may have been indirectly populated by collisions (green arrow and dotted lines in Fig. 2c).

The main difficulty that is met when modelling OP arises from the velocity dependence introduced by the combination of Doppler shifts and of the pump laser spectral profile. Atoms with a given velocity projection v_z along the OP light beam direction constitute a velocity class for which populations are coupled by OP rate equations where the contributions of absorption and emission of light are given by [8]:

$$\left.\frac{da_i}{dt}\right|_{OP} = \gamma \sum_{j=1}^{18} T_{ij} b_j + \sum_{j=1}^{18} \gamma_{ij}(b_j - a_i) \tag{13}$$

$$\left.\frac{db_j}{dt}\right|_{OP} = -\gamma b_j + \sum_{i=1}^{6} \gamma_{ij}(a_i - b_j), \tag{14}$$

where γ=1.022×10^7 s^{-1} is the radiative decay of the 2^3P state, T_{ij} is the transition matrix element, and γ_{ij} is the velocity-dependent OP rate of the transition $A_i \to B_j$. The OP rate γ_{ij} also depends on position $\mathbf{r}$ in the cell due to the transverse intensity profile of the pumping beam and to absorption along the beam path. For the case of a broadband laser of Gaussian linewidth L and intensity $I(\mathbf{r})$ tuned to the centre of the Doppler absorption profile, the OP rate at low pressure is given by:

$$\gamma_{ij}(v_z, \mathbf{r}) = \sqrt{\pi} I(\mathbf{r}) T_{ij} \frac{\alpha f}{m_e \omega_{ij} L} \exp\left[-\left(\frac{D v_z}{L \bar{v}}\right)^2\right], \tag{15}$$

where α is the fine structure constant, f=0.5391 is the oscillator strength of the transition, m_e is the electron mass, $\omega_{ij}/2\pi$ is the frequency of the $A_i \to B_j$ optical transition, $\bar{v}=\sqrt{2k_BT/M_3}$ is the

[3] A more realistic description of the depolarising effect of collisions in the 2^3P state (involving relevant cross sections for electronic angular momentum changing collisions [29], conservation of nuclear spin, and hyperfine recoupling following each collision) exceeds the scope of this report and will be given elsewhere. We use here the simple single-rate phenomenological approach used in previous models [8, 24]. It is expected to allow meaningful discussion of pressure effects on quantum efficiencies of C_8 and C_9 OP lines that address the same 2^3P_0 level.

most probable speed (T is the temperature, M_3 the atomic mass of ^{3}He), and $D=(\omega_{ij}/2\pi)\bar{v}/c$ is the associated Doppler width. In the model, the local OP light intensity $I(\mathbf{r})$ is computed from the incident laser power $W_{\rm las}$ and from the beam profile, taking into account the local atomic absorption in a self-consistent iterative way (forward and back-reflected intensities are added up in case of double pass).

Solving sets of rate equations for various velocity classes would be a difficult task since their populations are coupled by velocity-changing and by ME collisions, with ill-known collision rates. Instead, a coarse description with as few as two broad velocity classes is used to account for the Maxwell distribution of atomic velocities in the pumped gas: strongly pumped atoms, in the centre of the velocity distribution (with e.g. $v_z/\bar{v} \leq L/D$), and weakly pumped atoms, in the wings of the velocity distribution. This is a crude model, however it is sufficient to capture the essential features of velocity-selective OP effects and it usually provides reliable quantitative results.

In high magnetic field MEOP is performed on one of the intense $\mathrm{f}_2^{\pm}$ or $\mathrm{f}_4^{\mp}$ absorption lines (see Fig. 4) that combine 2 or 4 components with matrix elements close to 1 and thus provide much higher OP rates than low-field C_8 or C_9 lines at given pump intensity. In high pressure situations, three elements of the model are modified: full collisional redistribution in the 2^3P state simplifies Eqs. 13 and 14; the Doppler atomic absorption profile has to be replaced by a broader Voigt profile, which modifies the expression of the OP rate of Eq. 15 and decreases its value; fast collisional redistribution between velocity classes occurs in the 2^3S state and reduces the velocity dependent effects of OP, that can be neglected even if single-frequency laser pump is used. The MEOP model, with fewer phenomenological parameters ($\gamma_{\rm r}^{\rm P}$ and the partition of velocity classes are dropped), can thus be expected to provide more robust results.

2.5. Computed solutions of the complete rate equations at low field

In this section we quantitatively discuss the case of low pressure and low field OP that corresponds to currently operating MEOP-based gas polarisers. Specific features of high field situations [31] exceed the scope of this contribution and will be fully discussed elsewhere.

Coupled rate equations for the populations in the 2^3S and 2^3P state are obtained adding up the contributions of ME, OP, and relaxation (Eqs. 5, 11 and 12, 13 and 14). Characteristic rates of evolution for these two states are set by $\gamma_{\rm e}$ and γ, while the time evolution of the nuclear polarisation in the ground state (Eq. 18 below) involves rates typically six orders of magnitude smaller (e.g., $\gamma_{\rm e} n_{\rm m}/N_{\rm g}$ for ME). The coupled rate equations are thus solved adiabatically, by computing (quasi-) steady-state solutions in the 2^3S and 2^3P states assuming fixed values of all parameters, including the nuclear polarisation M in the ground state.

The effect of the competition between ME and OP is illustrated in Fig. 5 that displays the six populations of strongly pumped atoms in the 2^3S state for M=0 and M=0.5, in the absence of relaxation in the 2^3S state ($\gamma_{\rm r}^{\rm S}$=0). In the absence of OP spin-temperature populations are obtained, as expected (see Sec. 2.2): all populations are equal to 1/6 for M =0 and adjacent populations (for sublevels with $m_F' - m_F$=1) have a constant ratio for $M \neq 0$ (e.g., 3 for M=0.5). With the example of OP conditions chosen in Fig. 5 (double pass C_8 pumping with about 5 W/cm^2 of incident light intensity, yielding an effective pumping rate of order 4.8γ for the strongly pumped atoms), the addressed sublevel A_5 is efficiently depleted. All other sublevel populations are affected as well by the combined effects of 2^3P radiative decay paths in the OP cycle (see, e.g., Fig. 2c) and of hyperfine recoupling following ME collisions. In the 2^3P state, with the same MEOP conditions, the computed populations are b_{17}/a_5=0.77 for the radiatively populated level and b_j/b_{17}=0.023 for the other levels (that are populated by collisions).

Such modifications of populations with respect to the spin-temperature distribution strongly affect the way resonant light is absorbed. The pump light absorption, that depends on the difference of populations of the involved sublevels (a_5-b_{17} in the case of Fig. 5), significantly

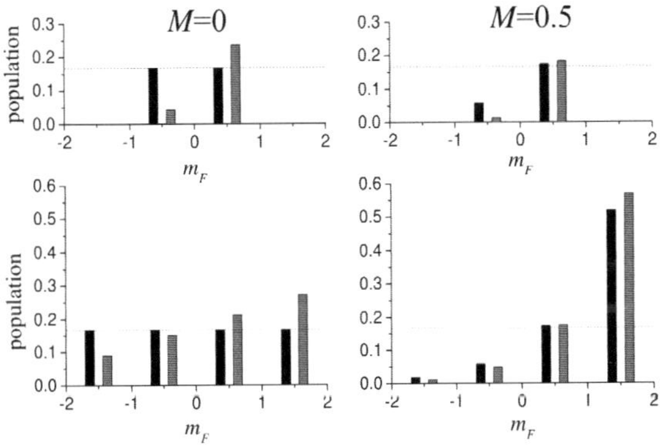

Figure 5. Computed populations in the 2^3S state for strongly pumped ^{3}He atoms at B=1 mT, for two values of M. Upper plots: populations a_5 and a_6 (F=1/2); lower plots: populations a_1 to a_4 (F=3/2). Black (left) bars: ST populations in the absence of OP. Red (right) bars: example of modified populations for strongly pumped atoms in a 1.33 mbar gas on the axis of the Gaussian pump beam (C_8 σ_+ pumping, $W_{\rm las}$=5 W, 1.33 cm fwhm diameter, 1.7 GHz fwhm spectral width; uniform $n_{\rm m}$= $10^{-6} \times N_{\rm g}$=3.2×10^{10} at/cm^3; $l_{\rm path}$ =2×30 cm).

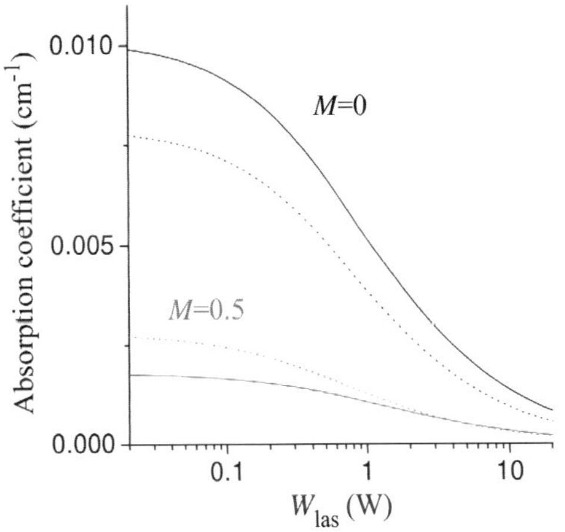

Figure 6. Overall pump absorption coefficients at M=0 (upper black curves) and M=0.5 (lower red curves) for C_8 (dotted lines) and C_9 (solid lines) OP, computed as a function of incident power $W_{\rm las}$ (same conditions as in Fig. 5).

decreases both with increasing M and with increasing pumping rate. This is illustrated in Fig. 6, in which the overall pump light absorption coefficient, -ln $T_{\rm pump}/l_{\rm path}$, is displayed as a function of pump power ($T_{\rm pump}$ is the computed pump transmittance for the absorbing gas, $l_{\rm path}$ is twice the cell length in the double-pass configuration). Computations are made for parameters corresponding to typical conditions of the experiments reported in the next section. Overall transmittance $T_{\rm pump}$ is computed by averaging over the transverse beam profile and over the two broad velocity classes. At fixed M, absorption significantly decreases at high light intensity: it is halved for $W_{\rm las}$=1 W in these conditions and would decrease more rapidly for a spectrally narrower laser or for a geometrically less expanded beam. Since it depends on $\gamma_{ij}/\gamma_{\rm e}$, the departure from the spin-temperature distribution under the influence of OP occurs at lower intensities for lower gas pressures. Fig. 6 also shows that the pump absorption decrease observed for increasing polarisation at fixed OP power is different for C_8 and C_9 lines.

An important consequence of the redistribution of populations by the intense OP light concerns the applicability of optical detection methods based on absorption measurements of a 1083 nm probe beam [24]. M can be accurately inferred from measurement of populations in the 2^3S state when a spin-temperature distribution prevails. An experimental study of the influence of OP-induced population changes on such optical methods is reported in a companion contribution in the same conference proceedings [32].

3. Photon efficiency and angular momentum budget

Numerical computation of all populations in the 2^3S and 2^3P states is indeed sufficient to fully characterise both the effect of OP on the electronic and nuclear orientations and on the absorption of light by the gas. However, it is useful for the sake of discussion to separately consider the angular momentum transfer from OP light to atomic variables, the effects of relaxation on the atomic angular momentum in each state, and the flow of angular momentum associated with ME collisions.

As was briefly mentioned in the discussion of the OP cycle (section 2.4 and Fig 2c), the net change of atomic angular momentum projection m_F upon the absorption and emission of a photon, that we call the OP photon efficiency (PE) η, depends on several factors. In all situations PEs depend on the magnetic field through ratios of transition intensities. For any transition addressing a single Zeeman sublevel of the 2^3S state, the PE does not depend on nuclear polarisation nor on OP intensity. This occurs for the C_8 line of ^{3}He and the D_0 line of ^{4}He at low field, and also for several high-field transitions. For the D_0 line, the PE additionally does not depend on pressure.

For C_8, η only depends on the relaxation rate in the 2^3P state:

$$\eta_{C8} = \frac{\gamma \eta^{K}_{C8} + \gamma^{P}_{r} \eta^{D}_{C8}}{\gamma + \gamma^{P}_{r}}, \tag{16}$$

where η^{K}_{C8}=0.896 is evaluated from transition intensities in the limiting case of low-pressure OP (discussed by Kastler [33]), i.e., when $\gamma^{P}_{r} \ll \gamma$ and η^{D}_{C8}=0.5 is evaluated in the opposite limiting case of depopulation OP (a high pressure regime discussed by Dehmelt [34]), i.e., when $\gamma^{P}_{r} \gg \gamma$.

For C_9 the overall PE is a weighted average of the efficiencies for absorption by the two 2^3S Zeeman sublevels addressed by the pump, hence it depends on the ratio of their populations that in turn depends both on M and on OP intensity. For instance, for σ_+ pump light the PE is higher (and larger than 1) for A_1 than for A_2. Furthermore, collisions in the 2^3P state enhance the PE in this case by increasing the final m_F value from -1/2 to 0 in the 2^3P state. This is illustrated in Fig 7 for the M-dependence of η_{C9} at low pump intensity. With the $3:1$ ratio (at B=0) of transition intensities from A_1 and from A_2, the PE is higher (and enhanced by collisional redistribution in the 2^3P state) at low M where absorption by A_1 dominates. The situation is reversed above M=0.5, where a_2/a_1=3 just compensates for the ratio of transition probabilities. Fig. 8 displays the power dependence of η_{C9} at fixed polarisations M=0 and M=0.5.

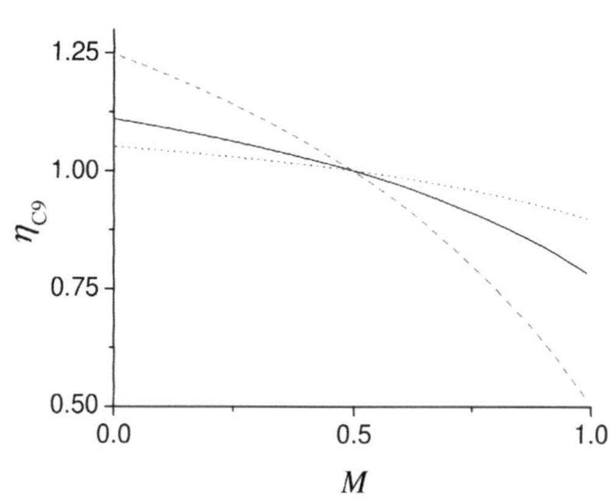

Figure 7. Computed M-dependence of the PE η_{C9} in the limit of low-intensity OP on the C_9 line. Red (dashed) line: η^{D}_{C9}, high pressure limit; black (solid) line: 1.33 mbar, intermediate collisional regime; blue (dotted) line: η^{K}_{C9}, low pressure limit.

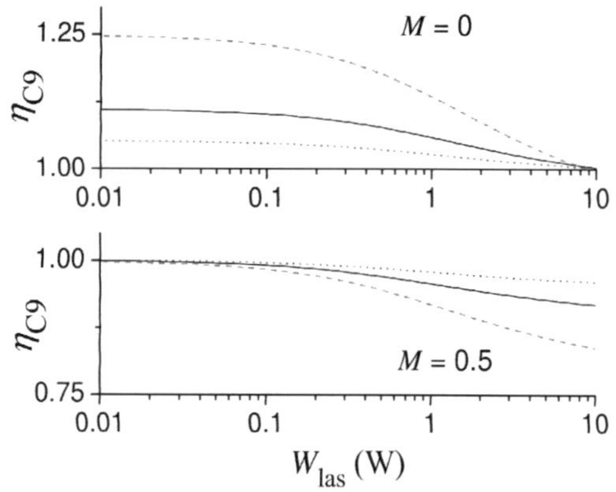

Figure 8. Computed power dependence of the PE of the C_9 line at M=0 (upper graph) and M=0.5 (lower graph) for the same three collisional regimes as in Fig. 7. η_{C9} is here the spatial average of the PE in the cell, computed for the same pump beam parameters as in Fig. 5.

Journal of Physics: Conference Series **294** (2011) 012002 doi:10.1088/1742-6596/294/1/012002

In high field the $f_2^{\pm}$ and OP lines $f_4^{\pm}$ consist, like C_9, of several components having different individual PEs. As a result their overall PEs also depend on M and on OP light intensity. They are additionally sensitive to laser light detuning from the centre of the composite pump line. For instance, at M=0 and low pump intensity the PEs are computed, and have been checked [31, 35], to be of order 1 for the $f_2^{\pm}$ lines and 0.5 for the $f_4^{\pm}$ lines. The PEs can be significantly increased by detuning the pump laser away from the less efficient component(s). This is also the case at finite M and for any pump power, which accounts for the higher OP performance systematically obtained in experiments with an appropriately detuned pump laser [31].

As regards the ground state, the contribution of ME collisions to the rate equation describing the evolution of M is obtained by computing $\mathrm{Tr}_n\, \rho_g I_z$ using Eqs. 3 and 4:

$$\left.\frac{dM}{dt}\right|_{\mathrm{ME}} = \gamma_{\mathrm{e}} \int_{\mathrm{cell}} \frac{d^3\mathbf{r}}{V_{\mathrm{cell}}} \frac{n_{\mathrm{m}}}{N_{\mathrm{g}}} \left(\langle I \rangle_{\mathrm{m}} - M \right), \quad \text{where} \quad \langle I \rangle_{\mathrm{m}} = \sum_{k=1}^{6} L_k a_k. \tag{17}$$

The nuclear polarisation is uniform in the ground state since relevant diffusion rates in a low pressure gas are usually much larger than typical rates of evolution for M. However both the nuclear polarisation $\langle I \rangle_{\mathrm{m}}$ and the density n_{m} of atoms in the metastable state can strongly vary over the volume of the cell, hence Eq. 17 involves a spatial average over the cell volume V_{cell}. Using (L_k)=$(-1,\, L_-,\, L_+,\, 1,\, -L_-,\, -L_+)$,[4] where the field-dependent parameters $L_\pm$=$\cos^2\theta_\pm - \sin^2\theta_\pm$ can be evaluated from the 2^3S state mixing parameters θ_- and θ_+ [24], the nuclear polarisation $\langle I \rangle_{\mathrm{m}}$ in the metastable state is directly derived from the computed populations. For B=0, the coefficients are $L_\pm$=$\pm 1/3$. In high fields they weakly differ from 1, which reduces the flow rate of angular momentum: a large number of ME collisions is required to transfer a unit of angular momentum between ground state and 2^3S state atoms [24, 35].

All other processes directly acting on the nuclear polarisation in the ground state tend to drive it to its (quasi-) null thermal equilibrium value with a relaxation rate Γ_{g}. The total rate equation for the nuclear polarisation of ground state atoms is thus given by:

$$\frac{dM}{dt} = -\Gamma_{\mathrm{g}} M + \left.\frac{dM}{dt}\right|_{\mathrm{ME}}. \tag{18}$$

The skewing of the distribution of populations enforced by OP (see Fig. 5) provides the driving term in the build-up of nuclear polarisation from an initially unpolarised state. Conversely, if OP is interrupted in a polarised gas, relaxation in the 2^3S state induces a loss of angular momentum that has to flow from the ground state reservoir of nuclear angular momentum. At low nuclear polarisation this loss is known to introduce for the ground state polarisation an additional decay with a rate Γ_{ME}^0 such that [19]:

$$\left.\frac{dM}{dt}\right|_{\mathrm{ME}}^{(W_{\mathrm{las}}=0,\ M\ll 1)} = -\Gamma_{\mathrm{ME}}^0 M, \quad \text{with} \quad \Gamma_{\mathrm{ME}}^0 = \frac{11}{3} \int_{\mathrm{cell}} \frac{d^3\mathbf{r}}{V_{\mathrm{cell}}} \frac{n_{\mathrm{m}}}{N_{\mathrm{g}}} \gamma_{\mathrm{r}}^{\mathrm{S}}. \tag{19}$$

For an arbitrary polarisation M we introduce the factor f_Γ defined by:

$$\left.\frac{dM}{dt}\right|_{\mathrm{ME}}^{(W_{\mathrm{las}}=0)} = -f_\Gamma \Gamma_{\mathrm{ME}}^0 M. \tag{20}$$

Polarisation decay in the plasma in the absence of OP is then described by:

$$\frac{dM}{dt}^{(W_{\mathrm{las}}=0)} = -\Gamma^{\mathrm{decay}} M, \quad \text{with} \quad \Gamma^{\mathrm{decay}} = \Gamma_{\mathrm{g}} + f_\Gamma \Gamma_{\mathrm{ME}}^0. \tag{21}$$

[4] This expression of the coefficients in (L_k) is valid up to B=0.1619 T; the same permutation of indices resulting from level crossings as in Eqs. 8 and 9 must be applied for higher field strengths.

Journal of Physics: Conference Series **294** (2011) 012002 doi:10.1088/1742-6596/294/1/012002

Computed f_Γ is found to be insensitive to plasma parameters (n_m, N_g, $\gamma_\mathrm{r}^\mathrm{S}$) and to the magnetic field strength over a wide range of conditions. It universally decreases with polarisation and has the M^2 dependence plotted in Fig. 9. The impact of this variation is discussed in Sect. 4.2 for a typical experimental polarisation decay.

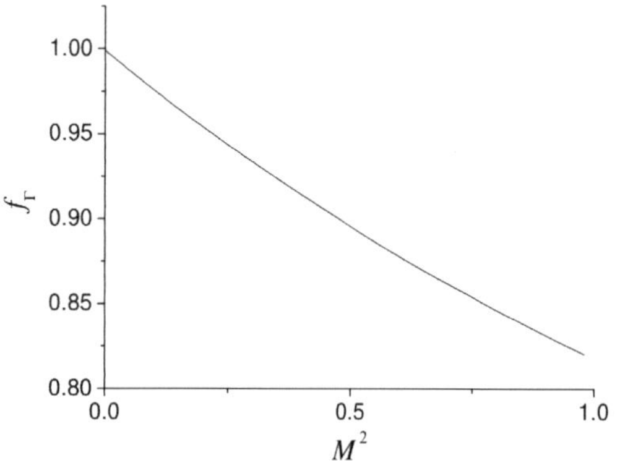

Figure 9. Computed decay rate factor f_Γ in Eq. 20characterising the deviation of ME-induced decay of M from the small M regime of Eq. 19. The empirical fit f_Γ=0.999−0.2340M^2+0.0539M^4 is accurate to better than 4×10^{-4}.

When OP is applied and losses occur in the 2^{3}S state due to relaxation ($\gamma_\mathrm{r}^\mathrm{S}\neq 0$), part of the angular momentum deposited by the OP cycles is dissipated and the remainder is transferred to the ground state. When a single sublevel is addressed by the OP (hence the PE is a constant) one can write a formula that conveniently accounts for this polarisation loss at finite M. For instance, for OP on the C$_8$ transition in low field the ME contribution (Eq.17) is written as:

$$\left.\frac{dM}{dt}\right|_{\mathrm{ME}} = 2\eta_{\mathrm{C8}}\frac{(1-T_{\mathrm{pump}})\,W_{\mathrm{las}}}{N_\mathrm{g}V_{\mathrm{cell}}\hbar\omega_{ij}} - f'_\Gamma\Gamma^0_{\mathrm{ME}}M. \tag{22}$$

The first term on the right hand side of Eq. 22 corresponds to angular momentum conservation in the absence of relaxation ($\gamma_\mathrm{r}^\mathrm{S}$=0, hence Γ^0_{ME}=0): the number of flipped nuclei per unit time, $(N_\mathrm{g}V_{\mathrm{cell}}/2)\ dM/dt|_{\mathrm{ME}}$, is just equal to the product of the PE, η_{C8}, by the number of absorbed photons per unit time, $(1-T_{\mathrm{pump}})\,I_{\mathrm{OP}}/\hbar\omega_{ij}$. The choice of the form of the second term results from the fact that, for any distribution of populations resulting from the competition of ME, OP, and relaxation, the computed angular momentum loss in the 2^{3}S state is very close, over the whole range of experimental parameters, to the one contributing to polarisation decay in the absence of OP (Eq. 20). Neglecting the difference between f'_Γ and f_Γ leads to a useful alternative form of Eq. 18:

$$\frac{dM}{dt} = -\Gamma_\mathrm{R}M + 2\eta_{\mathrm{C8}}\frac{(1-T_{\mathrm{pump}})\,W_{\mathrm{las}}}{N_\mathrm{g}V_{\mathrm{cell}}\hbar\omega_{ij}}, \quad \text{where } \Gamma_\mathrm{R} = \Gamma_\mathrm{g} + f_\Gamma\Gamma^0_{\mathrm{ME}}. \tag{23}$$

The total polarisation loss rate, Γ_R, is the sum of the decay rate Γ_g (corresponding to relaxation processes directly acting on atoms in the ground state) and of the additional decay rate introduced by ME (Eq. 20, resulting from relaxation of populations in the 2^{3}S state).

At null polarisation the total loss term in Eq. 23 vanishes and the photon efficiency η_{C8} (that does not depend on M) is very simply related to the growth rate of polarisation:

$$\eta_{\mathrm{C8}} = \frac{1}{2}\frac{dM}{dt}^{(M=0)}\frac{N_\mathrm{g}V_{\mathrm{cell}}\hbar\omega_{ij}}{\left(1-T^{(M=0)}_{\mathrm{pump}}\right)W_{\mathrm{las}}}. \tag{24}$$

Eq. 23 can thus be rewritten so as to relate the total loss rate Γ_R to directly measurable quantities:

$$\Gamma_\mathrm{R}M = \frac{1-T_{\mathrm{pump}}}{1-T^{(M=0)}_{\mathrm{pump}}}\frac{dM}{dt}^{(M=0)} - \frac{dM}{dt}. \tag{25}$$

This takes a simple form at steady-state polarisation M^{eq}, where the total loss term exactly balances the angular momentum deposited by the OP photons:

$$\Gamma_{\mathrm{R}} M^{\mathrm{eq}} = \frac{1 - T_{\mathrm{pump}}^{(M=M^{\mathrm{eq}})}}{1 - T_{\mathrm{pump}}^{(M=0)}} \frac{dM}{dt}^{(M=0)} . \quad (26)$$

Given the various parameters appearing in the rate equations, the model indeed allows to directly compute the ME contribution to the time evolution of M (Eq. 17), hence the kinetics of polarisation growth (Eq. 18) and the steady-state value M^{eq}, if the loss rate Γ_{g} is known. It also allows to compute pump beam transmission and photon efficiencies that can be directly compared to measurable quantities, for instance to ascribe a value to the pressure-dependent relaxation rate $\gamma_{\mathrm{r}}^{\mathrm{P}}$. Angular momentum budget considerations provide much less insight on the detailed mechanism of MEOP but they lead to very robust predictions about the link between pump light absorption and polarisation kinetics (independent of specific features of this MEOP model, such as the coarse treatment of velocity-dependent OP rates).

4. Experimental results

Several experiments devoted to the quantitative study of ^{3}He MEOP under various non-standard conditions have been performed in the recent years. We report here on selected results, mainly regarding nuclear relaxation, obtained using an intense pumping light in standard MEOP conditions (low gas pressure and low magnetic field) [30] and we compare them to results obtained combining high gas pressure and high magnetic field [13, 31, 35].

4.1. Experimental setup and polarisation measurement at low field

A dedicated setup, schematically described in Fig. 10, has been constructed to characterise MEOP kinetics and performance over a wide range of conditions encompassing the usual operating parameters of gas polarisers. An end-compensated solenoid provides a uniform

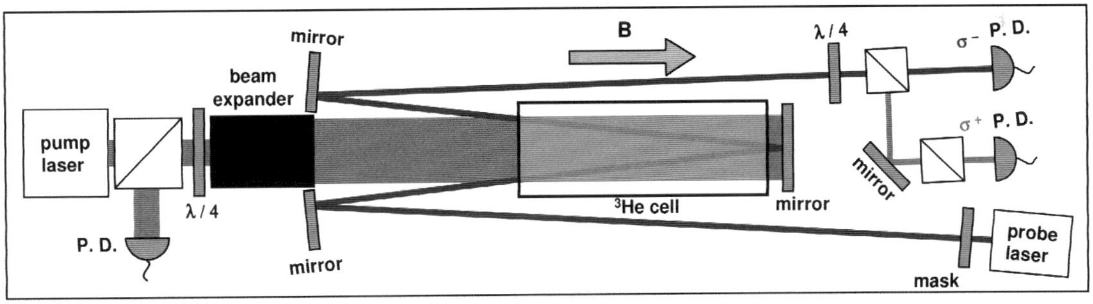

Figure 10. Schematic of the set-up used for low-field MEOP studies. Pump and probe laser lights are delivered to the experiment by single-mode fibres. The Gaussian pump beam is collimated by a f=7 cm lens, circularly polarised by a polarising beam splitting cube and a quarterwave plate ($\lambda/4$), and expanded to a diameter of 1.4 cm fwhm. The weak probe beam is split and polarised by a mask to obtain two vertically separated adjacent beams (this is not visible in this top view). The probe beams, with opposite circular polarisations, are separated by a circular analyser. Photodiodes (P.D.) are used to monitor the transmitted pump and probe beam powers after a double pass through the cell .

magnetic field (B=0–30 mT, mean $\Delta B/B$=50 ppm/cm) over the volume of the experimental cells. These are sealed cylindrical quartz cells (30 cm long, 6 cm in diameter) with optically polished windows, filled with pure ^{3}He gas after a standard cleaning protocol. We report here on results of studies performed at B=1 mT for three gas pressures: 0.63, 1.19, and 2.45 mbar. A rf discharge is sustained in the cell using two sets of alternating wire electrodes connected to a voltage transformer tuned to the rf frequency (of order 2 MHz). Partial amplitude modulation of the rf power (around 70 Hz) and lock-in detection are used to improve the sensitivity of the absorption measurements. For most experiments, a broadband 1083 nm fibre laser provides the OP pump power (Keopsys, 5 Watt model, 1.7 GHz fwhm). For enhanced OP efficiency the light is back-reflected for a second pass through the cell. After deflection by the circular polariser, the transmitted light is monitored for measurements of the absorbed pump power. A longitudinal optical detection method is used to measure the nuclear polarisation in the gas during experiments [32]. It is based on absorption measurements of two weak probe beams (provided by a 1083 nm DBR diode) arranged in a W-shaped path for convenience (see Fig. 10), having complementary σ_+ and σ_- circular polarisations separately monitored using two photodiodes.

Examples of demodulated and processed signals obtained from the recordings of the photodiode output voltages are displayed in figure 11 for C_8-σ_+ OP experiments where short periods of polarisation build-up alternate with periods with a blocked pump. The upper graphs

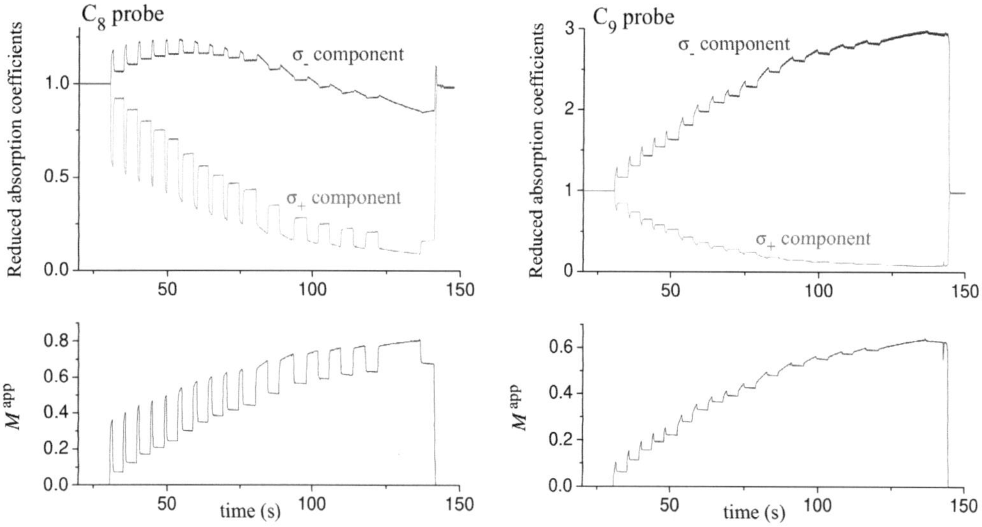

Figure 11. Two processed time recordings obtained from probe photodiode voltages for the 1.19 mbar cell. Left (resp., right) graphs: the probe laser is tuned to C_8 (resp., C_9). Both recordings are made with the OP beam (σ_+-polarised, tuned to C_8, $W_{\rm las}$=1.55 W) alternatively applied and blocked. Upper graphs: reduced absorption coefficients are computed from demodulated photodiode voltages (see Sec. 3.5 in [32]). Lower graphs: time evolutions of the apparent polarisations $M^{\rm app}$ are computed from the signals in the upper graphs using Eqs. 3 and 5 of [32]. When the pump is blocked $M^{\rm app}$ is equal to the ground state polarisation M. When the pump is applied the instant increase in $M^{\rm app}$ reveals the departure from a spin-temperature situation and the overpolarisation of the 2^3S state. This increase is larger when the pump and the probe address the same Zeeman sublevels (left versus right bottom graphs).

display reduced absorption coefficients (-ln T/-ln $T^{(M=0)}$) of the probe beams. Initially M=0 and both components are equally absorbed. Then, during the polarisation build-up, the absorption of the σ_+-polarised beam steadily decreases because of the decrease of the populations of sublevels with m_F<0 (a_1, a_2, and a_5). Additional transient decrease of absorption is observed whenever the pump light is applied, which results from the strong distortion of the distribution of populations by intense OP (see Fig. 5). The sublevel A_5 is the most depleted and the 2^3P sublevel B_{17} is the most populated by the pump light. Probe light absorption is thus strongly decreased for the σ_+ component of the C_8-tuned probe beam, and much less for the σ_+ component of the C_9-tuned probe beam. Conversely, OP light and growing polarisation both tend to overpopulate sublevels with $m_F>0$ and thus to enhance the absorption of the σ_--polarised probe beam light. After reaching a maximum, absorption of the C_8 probe light would vanish at $M=1$ while absorption of the C_9 probe light would be enhanced by a factor of 4.5 compared to $M=0$.

These data provide a clear manifestation of the spin-temperature enforced distribution of populations in the absence of OP and of the overpolarisation induced by the pumping light in the 2^3S state. Similar recordings have also been performed with the probe laser detuned from resonance in order to address atoms of different velocity classes. This can be used to obtain quantitative information on velocity-selective OP effects and to evaluate the relevance and the limitations of the simple two-class model introduced in Section 2.4.

The lower graphs in Fig. 11 display the apparent polarisations M^{app} computed from the ratios of reduced absorption signals by assuming a spin-temperature distribution of populations (Eqs. 8 and 9). They show sawtooth patterns, with rising M^{app} (always overestimating the growing M) when the OP light is applied and decaying M when the OP beam is blocked. Such sets of data are used to accurately relate M^{app} to M for given OP conditions and choice of probe line. They are also used to test the MEOP model. Comparison to the computed apparent polarisations is quite satisfactory, with for instance a difference between computed and measured M^{app} values that do not exceed 0.015 for these experiments.

4.2. Characterisation of build-up and decay of polarisation

A standard MEOP experiment consists in monitoring transmitted pump and probe beam powers for selected rf intensity for the discharge, OP power and line, and probe laser line (C_8 or C_9). A typical recording of polarisation build-up and decay is displayed in Fig. 12.

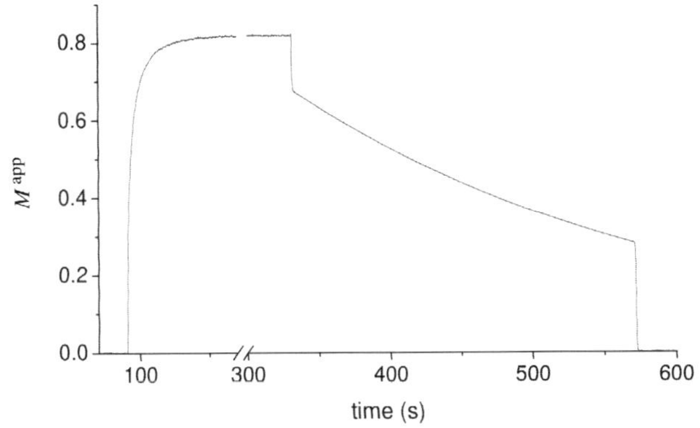

Figure 12. Time evolution of the apparent polarisation M^{app} in the 1.19 mbar cell, monitored using a C_8-tuned probe. OP is applied from t=91 s to t=330 s (C_8 line, W_{las}=1.55 W). At t=570 s, the remaining polarisation is suddenly destroyed by enhanced nuclear relaxation (obtained by placing a small magnet close to the cell).

The build-up part is best discussed using Fig. 13a, in which the true polarisation M is inferred from M^{app} by applying the empirical correction function determined from the dedicated measurement of M^{app}-M in the same OP conditions (Fig. 11). The difference of M to the equilibrium polarisation M^{eq}=0.67 that is reached in steady-state in this experiment is plotted in semi-log scale (Fig. 13a). Departure from a straight line reveals a non-exponential approach of M towards M^{eq}, systematically observed at low magnetic fields. The changing slope of the curve corresponds to a build-up rate $\Gamma_{build}=(dM/dt)/(M^{eq}\text{-}M)$ that decreases while M grows towards M^{eq}. Fig. 13b displays values of dM/dt inferred from a running fit of the polarisation data. It is not a straight line and the local slope (Γ_{build}, data not shown) is found to decrease

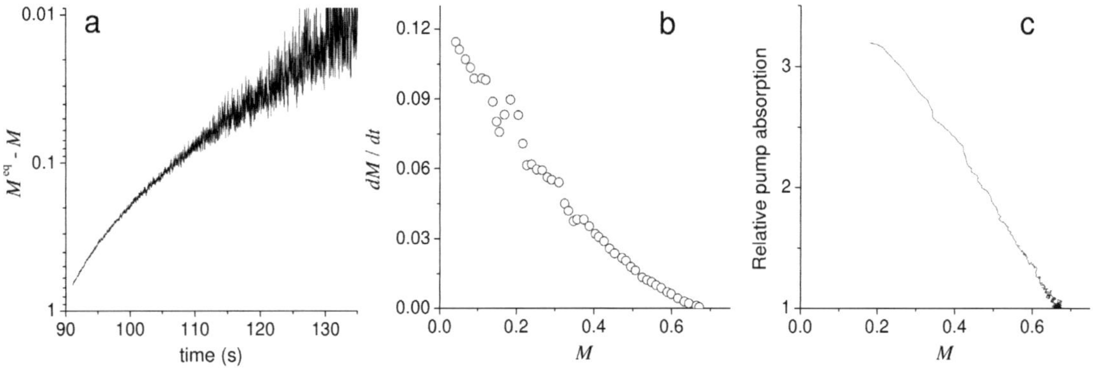

Figure 13. Analysis of polarisation build-up for the experiment displayed in Fig. 12. **a:** The difference of the true polarisation M to its steady-state value $M^{\rm eq}$ is plotted in a reversed semi-log scale as a function of time. The build-up is clearly non-exponential. **b:** The time derivative of M, inferred from local exponential fits to data, is plotted as a function of M. Extrapolation at M=0 provides $dM/dt^{(M=0)}$=0.125 s^{-1} **c:** Relative pump light absorption is plotted as a function of M. For convenience, the data are here normalised to the value reached at M=$M^{\rm eq}$, where a transient switching off of the rf discharge (during the axis break in Fig. 12) is used to measure absolute pump and probe transmittances.

with M quite linearly, from 0.19 s^{-1} at M=0 to 0.06 s^{-1} at M=$M^{\rm eq}$. The absorbed pump power, that is simultaneously recorded in the experiment, is plotted in Fig. 13c as a function of the polarisation. Due to the time constant set for high signal to noise ratio in this experiment, no meaningful absorption value is obtained during the first second of pumping. Still, pump absorption clearly decreases by a factor of order 4 when M increases from 0 to $M^{\rm eq}$.

The decay part of the recording in Fig. 12 is best discussed using Fig. 14a, where decaying polarisation is plotted in semi-log scale. When the pump is blocked (after t=330 s), $M^{\rm app}$=M can be accurately fit by an exponential function decaying to 0 (with a rms difference below 3×10^{-4}). This is a generic feature of polarisation decays in rf discharges, for which constant $\Gamma^{\rm decay}$ are obtained in all operating conditions and over a wide range of measured decay rates (10^{-4}–10^{-1} s^{-1}).

During decay a spin-temperature distribution of populations prevails and the populations are precisely known (Eqs. 8 and 9). The measured transmitted powers can thus be used to obtain the value of the metastable number density $n_{\rm m}$ (integrated along the probe path). Absolute transmittances are obtained only when the rf discharge is turned off but relative changes are reliably derived from demodulated signals at all times. This is used to obtain the data in Fig. 14b where the 2^3S number density $n_{\rm m}$ exhibits an approximate M^2-dependence during polarisation decay. The significant change in $n_{\rm m}$ (-15% when M decays from 0.67 to 0), that may for instance be attributed to the inhibiting effect of polarisation on Penning ionising collisions, induces an identical change of $\Gamma^0_{\rm ME}$ (Eq. 19) that is not compensated by the variation of the factor f_Γ (Fig. 9) in Eq. 21. The change in $n_{\rm m}$ also indicates that the plasma conditions are modified and suggests that $\Gamma_{\rm g}$ may vary with polarisation. Altogether, since $\Gamma^{\rm decay}$=$\Gamma_{\rm g}+f_\Gamma\Gamma^0_{\rm ME}$, the precisely exponential character of decays recorded in very different experimental situations is puzzling and can certainly not be attributed to an accidental compensation for these independently varying quantities.

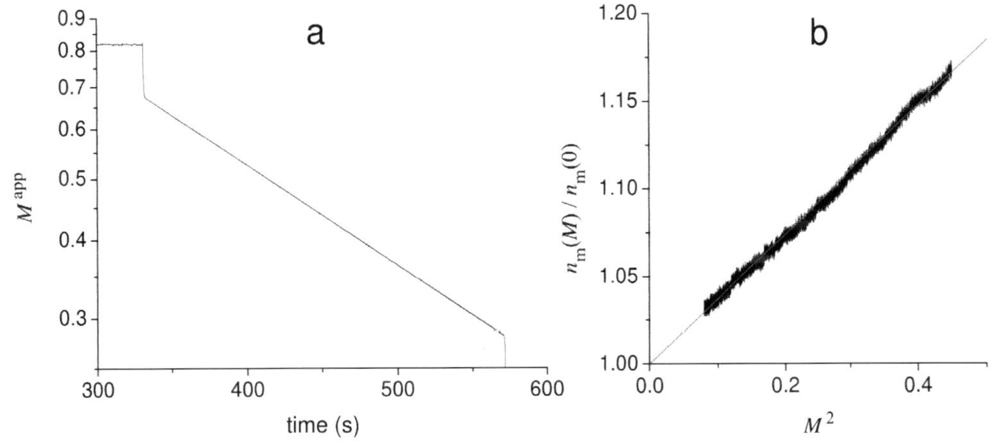

Figure 14. Analysis of polarisation decay for the experiment displayed in Fig. 12. **a:** M^{app}, that reduces to M when the pump is blocked (after t=330 s), is plotted in a semi-log scale as a function of time. The decay is accurately exponential, with a rate Γ^{decay}=3.667×10^{-3} s^{-1}. Extrapolation to t=330 s yields M^{eq}=0.67. **b:** The relative change of the number density n_m of atoms in the 2^3S state is plotted as a function of M^2 for data taken during polarisation decay. The dotted line is a guide for the eye (slope: 0.37).

4.3. Steady-state polarisation and ground state nuclear relaxation

Steady-state nuclear polarisation M^{eq} is asymptotically reached when the ME contribution to the time evolution of M (Eq. 17) is just balanced by the direct relaxation term $-\Gamma_g M$ in Eq. 18. Figure 15 displays computed values of this ME contribution for C_8 and C_9 OP and conditions corresponding to the recording of Fig. 12 (W_{las}=1.55 W, p=1.19 mbar, probe transmittance T=0.48 at M=0, Γ^{decay}=3.667×10^{-3} s^{-1}). For these calculations, 2^3S and ground

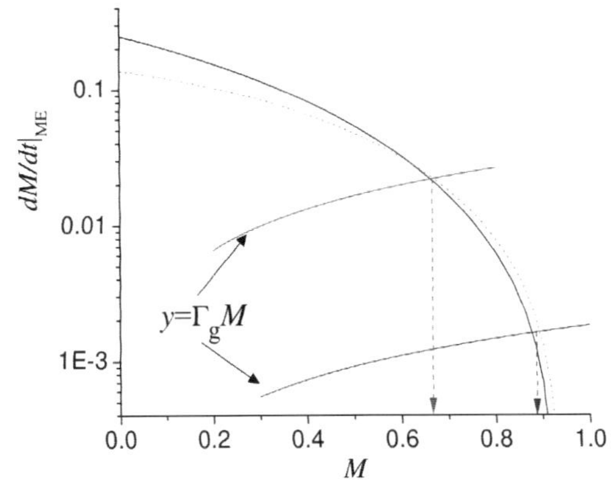

Figure 15. ME contributions to the evolution of M (Eq. 17) are computed as a function of M for OP on C_9 (solid black line) and C_8 (dotted red line) lines and plasma conditions specified in the text. The semi-log scale is used for convenient comparison of small values at large M. In this representation, the linear functions y=$\Gamma_g M$ are not straight lines (upper curve: Γ_g=0.033 s^{-1}; lower curve: Γ_g=1.83×10^{-3} s^{-1}). The vertical arrows point at the corresponding steady-state values M^{eq} on the M-axis.

state relaxations are assumed to equally contribute to the measured polarisation decay of Fig. 14: Γ^0_{ME}= Γ_g=$\Gamma^{decay}/2$. If one assumes, as usual, that nuclear relaxation of ground state atoms is not affected by the presence of the OP light, M^{eq} is graphically found to be of order 0.9 in Fig. 15 (at the crossing of the computed ME contribution with the lower of the two curves y=$\Gamma_g M$). This value is found to be almost independent (±0.03 at most) of the chosen 2^3S relaxation rate, provided that $\Gamma^0_{ME}+\Gamma_g=\Gamma^{decay}$. It is much higher than the steady-state polarisation measured in the 1.19 mbar cell, M^{eq}=0.67. This experimental value can be obtained at steady-state using the MEOP model only by assuming that Γ_g is much larger (here, 0.033 s^{-1}, corresponding to the upper of the two curves $y=\Gamma_g M$) with the OP beam than without it .

Discrepancies are systematically obtained between experimental results and M^{eq} values computed with the MEOP model assuming that ground state relaxation is not affected by OP This is illustrated in Fig. 16 for a series of measurements in the other two cells used in this study. Computed results weakly depend on input parameters such as the spatial distribution of

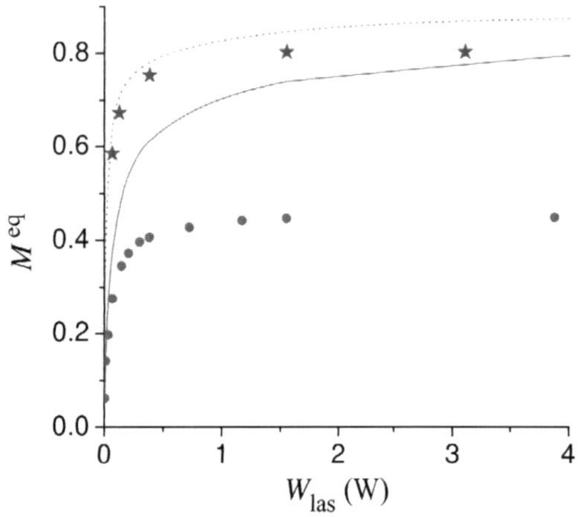

Figure 16. Measured (symbols) and computed (lines) values of the steady-state polarisations M^{eq} are plotted as a function of the incident pump power W_{las} for OP on C_8 in the 0.63 mbar cell (green stars and dotted line) and on C_9 in the 2.45 mbar cell (red circles and solid line). Computations are made for the parameters of the experiments (n_m=1.1 and 4.2×10^{16} at/m^3 and $\Gamma_R=\Gamma^{decay}$= 2.9 and 13×10^{-3} s^{-1} at low and high pressure, corresponding to a weak and a rather intense discharge, respectively).

2^{3}S atoms in the cells, the relaxation rates γ_r^S and γ_r^P, the OP laser spectral and geometrical profiles, etc. Assuming a significant, power-dependent increase of Γ_g during OP is the only way to reproduce experimental M^{eq} data with the MEOP model.

In the case of OP on the C_8 line, the angular momentum budget approach of Sec. 3 relating the total loss rate Γ_R introduced in Eq. 23 to directly measured quantities provides a model-independent and parameter-free approach to evaluate Γ_R during polarisation build-up (using Eq. 25) and at steady-state polarisation (using Eq. 26). When applied to the recording of Fig. 12 (C_8 pumping), Eq. 26 yields Γ_R=0.047 s^{-1} at $M=M^{eq}$. Given the uncertainties on the extrapolations at M=0 in Figs. 13b and 13c, this is in good agreement with the result obtained using the MEOP model for the Fig. 15 experiment where an OP-enhanced relaxation rate Γ_R=0.0347 s^{-1} is required to account for the experimental observations. Here both inferred values of Γ_R are about 10 times higher than Γ^{decay}.

When the angular momentum approach is applied to a full build-up recording to directly infer total loss rates from the measured pump transmittances and M growth rates (using Eq. 25) a time-dependent series of values for Γ_R is obtained that includes the value derived at M^{eq} as an end-point. Remarkably consistent sets of data are obtained when they are plotted as a function of the absorbed pump powers, as displayed in Fig. 17 that compiles the results from 13 C_8-pumping experiments performed at various incident pump powers in the three cells. At high pump powers all data collapse to a common line with Γ_R proportional to the absorbed power and OP-enhanced relaxation rates Γ_R reach values that are up to two orders of magnitude larger than Γ^{decay}. At very low pump powers Γ_R would indeed recover the respective Γ^{decay} values measured in each cell for the chosen weak rf discharge level.

The full results of this study, performed for various discharge intensities and in fields up to 30 mT, will be reported elsewhere. They show a large and consistent increase of the ground state nuclear relaxation with the absorbed OP power. It is also worth comparing this with preliminary evidence of the same behaviour that has been obtained in MEOP studies at higher

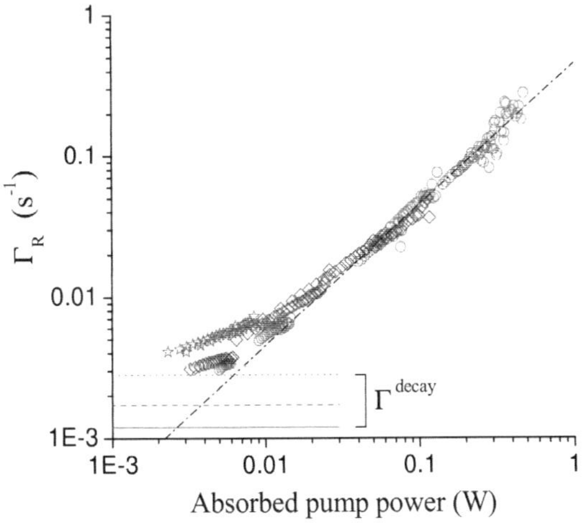

Figure 17. Examples of experimental total polarisation loss rates Γ_R obtained from polarisation build-ups (using Eq. 25) are plotted versus the absorbed OP power (C_8 pumping). The horizontal lines correspond to the measured Γ^{decay} values. Green stars and dotted line: 0.63 mbar; blue diamonds and dashed line: 1.19 mbar; red circles and solid line: 2.45 mbar. The dash-dotted line (y=x/2.2) is a guide for the eye.

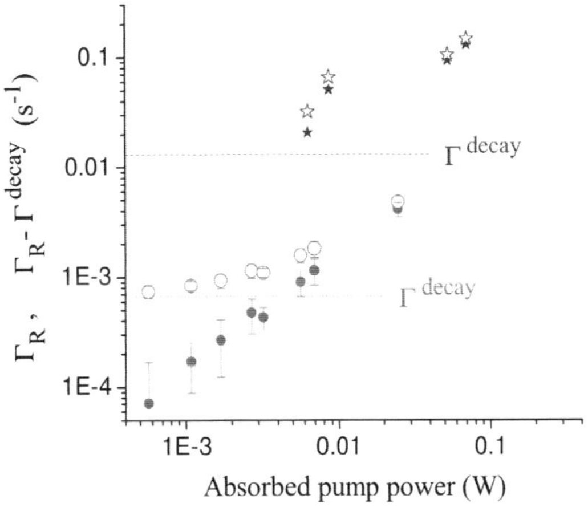

Figure 18. Total polarisation loss rates Γ_R (open symbols) for various absorbed OP powers in a 32 mbar cell, reproduced from Fig. 6.21 of [31]. The horizontal lines correspond to the measured Γ^{decay} values. The filled symbols are the differences Γ_R-Γ^{decay} plotted in the original figure. Blue stars and dashed line: B=3 mT, C_9 pump; red circles and dotted line: B=1.5 T, f_2^- pump.

gas pressures and applied fields [13, 35]. Figure 18, adapted from Fig. 6.21 of [31], displays Γ_R values inferred from the comparison of measured and computed M^{eq} data in a 32 mbar 5 cm×5 cm cell (using the MEOP model as shown in Fig. 15). At this pressure, MEOP is quite inefficient at low field (M^{eq}=0.11 with 2 W OP on the C_9 line at 3 mT) where rather large decay rates are measured (here, Γ^{decay}=13×10^{-3} s^{-1}), even for weak rf discharges. Still, OP-enhanced relaxation is observed and results are consistent with those presented in Fig. 17 if the appropriate cell length scaling is applied. Another striking feature is that high-field MEOP, which operates quite efficiently (M^{eq}=0.56 with 0.5 W OP at 1.5 T in the same cell) and with reduced decay rates (here, Γ^{decay}=0.67×10^{-3} s^{-1}), also gives rise to a strong OP-light-enhanced relaxation. This is currently being investigated at still higher pressures and magnetic field in Cracow [14, 15, 36].

5. Concluding remarks

We have described the main features of a comprehensive model that has been developed to account for the detailed features of MEOP in ^{3}He and in isotopic mixtures over a broad range of experimental conditions regarding field strength, gas pressure, pump laser intensity, and spectral characteristics. Selected results have been presented to illustrate the satisfactory agreement obtained bewteen computed and experimental data. The model has a limited number of free parameters and can be used for quantitative predictions of MEOP performance in gas polarisers (polarisations, growth rates, and magnetisations, with or without gas flow in OP cells) when reliable input values are available.

The model is used as a unified frame to understand, for instance, the surprisingly good MEOP results obtained in very high magnetic fields in spite of the strong hyperfine decoupling that tends to impede the transfer of angular momentum from the absorbed polarised light to the ^{3}He nuclei. The same high photon efficiency (of order 1) is obtained at all field strengths, which is a key factor in a global angular momentum budget. Since lower nuclear relaxation rates are obtained when a high field is applied to a rf discharge, in particular at high gas pressures where ME collisional couplings are advantageously increased, this high photon efficiency provides high nuclear polarisations over a broader range of operating conditions, hence of potential applications of the MEOP technique.

Consistent experimental observations of strongly OP-enhanced polarisation loss rates, quantitatively inferred from the measured MEOP dynamics using the model or the angular momentum budget approach, provide solid evidence of a new nuclear relaxation process. The additional loss rates account for the surprising experimental fact that the achieved steady-state polarisations hardly increase when higher pump powers are used, in spite of higher pumping rates. Search for the physical origin of this light-induced relaxation that occurs at low and high field strengths is under way, with dedicated optical tools used to probe plasma changes during MEOP. Finding ways to alleviate this limiting relaxation problem could open the way to improved performance of the MEOP technique, both in terms of record polarisations and of increased throughput for gas polarisers.

References

[1] Colegrove F D, Schearer L D and Walters G K 1963 *Phys. Rev.* **132** 2561–72
[2] Schearer L D 1969 *Phys. Rev.* **188** 505
[3] Lefevre-Seguin V and Leduc M 1977 *Journal of Physics B: Atomic and Molecular Physics* **10** 2157–64
[4] Xia T, Morgan S W, Jau Y Y and Happer W 2010 *Phys. Rev. A* **81** 033419
[5] Bouchiat M A, Carver T R and Varnum C M 1960 *Phys. Rev. Lett.* **5** 373–5
[6] Happer W 1972 *Rev. Mod. Phys.* **44** 169–249
[7] Chen W C, Gentile T R, Walker T G and Babcock E 2007 *Phys. Rev. A* **75** 013416
[8] Nacher P and Leduc M 1985 *Journal de Physique* **46** 2057–73
[9] Mueller R 2001 *Physica B: Condensed Matter* **297** 277–81
[10] Gentile T, Hayden M and Barlow M 2003 *J. Opt. Soc. Am. B* **20** 2068–74
[11] Tastevin G, Grot S, Courtade E, Bordais S and Nacher P J 2004 *Appl. Phys. B: Lasers and Optics* **78** 145–56
[12] Nacher P J, Courtade E, Abboud M, Sinatra A, Tastevin G and Dohnalik T 2002 *Acta Physica Polonica B* **33** 2225–36
[13] Abboud M, Sinatra A, Maitre X, Tastevin G and Nacher P J 2004 *Europhysics Letters* **68** 480–6
[14] Nikiel A, Palasz T, Suchanek M, Abboud M, Sinatra A, Olejniczak Z, Dohnalik T, Tastevin G and Nacher P J 2007 *European Physical Journal - Special Topics* **144** 255–63 ISSN 1951-6401
[15] Nikiel A, Collier G, Glowacz B, Palasz T, Olejniczak Z, Weglarz W, Tastevin G, Nacher P-J, Dohnalik T *in preparation*
[16] Greenhow R C 1964 *Phys. Rev.* **136** A660–A2
[17] Daniels J, Timsit R, May A and Yuen V 1971 *Canadian Journal of Physics* **49** 517
[18] Daniels J and Timsit R 1971 *Canadian Journal of Physics* **49** 525
[19] Dupont-Roc J, Leduc M and Laloë F 1973 *Journal de Physique* **34** 961–76 and 977–87
[20] Leduc M, Nacher P J, Tastevin G and Courtade E 2000 *Hyperfine Interactions* **127** 443–9
[21] Wolf M 2004 *Erzeugung höchster 3He Kernspinpolarisation durch metastabiles optisches Pumpen* Ph.D. thesis Johannes Gutenberg-Universität Mainz URL `http://nbn-resolving.de/urn/resolver.pl?urn=urn:nbn:de:hebis:77-6554`
[22] Stoltz E, Meyerhoff M, Bigelow N, Leduc M, Nacher and Tastevin G 1996 *Appl. Phys. B: Lasers and Optics* **63** 629–33
[23] Larat C 1991 *Laser LNA de puissance, application au pompage optique de l'hélium-3 et des mélanges hélium-3/hélium-4* Ph.D. thesis UPMC-Paris6 URL `http://tel.archives-ouvertes.fr/tel-00011879`
[24] Courtade E, Marion F, Nacher P J, Tastevin G, Kiersnowski K and Dohnalik T 2002 *European Physical Journal D - Atomic, Molecular, Optical and Plasma Physics* **21** 25–55
[25] Pinard M and Laloë F 1980 *Journal de Physique* **41** 799–818
[26] Dupont-Roc J, Leduc M and Laloë F 1971 *Phys. Rev. Lett.* **27** 467–70
[27] Fitzsimmons W A, Lane N F and Walters G K 1968 *Phys. Rev.* **174** 193–200
[28] Schearer L D 1967 *Phys. Rev.* **160** 76–80
[29] Vrinceanu D, Kotochigova S and Sadeghpour H R 2004 *Phys. Rev. A* **69** 022714+

[30] Batz M 2011 *Metastability exchange optical pumping in ^{3}He gas up to 30 mT: Efficiency measurements and evidence of laser-induced nuclear relaxation* Ph.D. thesis UPMC - Paris 6 and Johannes Gutenberg-Universität Mainz to be published on URL `http://tel.archives-ouvertes.fr/`
[31] Abboud M 2005 *Pompage optique de l'hélium-3 à forte pression dans un champ magnétique de 1.5 Tesla* Thèse de doctorat Université Pierre et Marie Curie URL `http://tel.archives-ouvertes.fr/tel-00011099/`
[32] Talbot C, Batz M, Nacher P J and Tastevin G 2011 *IOP journal of physics conference series, this volume*
[33] Kastler A 1957 *J. Opt. Soc. Am.* **47** 460–5
[34] Dehmelt H G 1957 *Phys. Rev.* **105** 1487–9
[35] Abboud M, Sinatra A, Tastevin G, Nacher P and Maitre X 2005 *Laser Physics* **15** 475–9
[36] Suchanek K, Suchanek M, Nikiel A, Palasz T, Abboud M, Sinatra A, Nacher P J, Tastevin G, Olejniczak Z and Dohnalik T 2007 *European Physical Journal - Special Topics* **144** 67–74

Polarized ^{3}He cell development and application at NIST

W C Chen[1,2], T R Gentile[1*], C B Fu[1,3], S Watson[1], G L Jones[1,4], J W McIver[1,5], D R Rich[1,6]

[1] National Institute of Standards and Technology, Gaithersburg, Maryland, USA 20899
[2] University of Maryland, College Park, Maryland, USA 20742
[3] Indiana University, Bloomington, Indiana, USA 47408

E-mail: thomas.gentile@nist.gov

Abstract. Over a period of well over a decade, a large number and variety of polarized ^{3}He cells for neutron applications have been developed and tested at the U.S. National Institute of Standards and Technology (NIST). These cells have primarily been employed for spin-exchange optical pumping (SEOP), and applied to neutron scattering and fundamental neutron physics. We describe the procedures we have employed for producing these cells, as well as their characteristics and applications. Whereas our best results are for cells blown from boron-free aluminosilicate glass, we summarize results with a range of other methods. We discuss our recent work on SEOP cells for wide angle neutron polarization analysis.
Key words: polarized

1. Introduction

Production and application of polarized ^{3}He by spin-exchange optical pumping (SEOP)[1, 2] depends critically on the characteristics of the cells in which the optical pumping occurs. A relaxation time long compared to the slow pumping time constants typical for SEOP is a key requirement. Application to neutron spin filters (NSFs) put additional demands on cell fabrication. Although there is currently a thrust towards continuous optical pumping for neutron spin filters, remote optical pumping is often employed. In this case there is an additional, often stronger need for long relaxation time beyond that which is required for the SEOP process itself. To avoid the need for gas transfer and recycling, as well as to allow for future continuous operation, it is often desirable to directly polarize ^{3}He cells for use on the neutron beam line. In this case, the need to heat cells to typical temperatures of 170 °C for pure rubidium or 220 °C for Rb/K hybrid SEOP[3] presents additional issues for SEOP cells, in particular for relaxation phenomena[4, 5]. Neutron spin filter cells cannot contain any ^{10}B, further constraining the choice of acceptable glasses. These requirements have led to the use of boron-free aluminosilicate glass, which has low He permeability, high neutron transmission, and good alkali resistance. For well over a decade, a large number and variety of cells have been made at the National Institute of Standards and Technology (NIST) for the development and application of polarized ^{3}He. These

[4] now at Hamilton College, Clinton, New York
[5] now at Harvard University, Cambridge, Massachusets
[6] deceased

cells have primarily been for studies of SEOP and SEOP-based neutron spin filters, but have also included cells for metastability-exchange optical pumping (MEOP[6]). In Sec. 2 we present a detailed discussion of our procedures and experience in preparing these cells and in Sec. 3 we discuss apparatus and results for characterization of cells. In Sec. 4 we survey the wide range of behavior of relaxation times and discuss some special cell constructions. The status of our most recent goal of fabricating cells for wide-angle polarization analysis[7] is presented in Sec. 5 and we summarize the paper in Sec. 6. The Appendix contains detailed lists of a large range of cells and the applications they have made possible.

2. Cell Fabrication

In this section we provide the detailed procedure we have employed to produce ^{3}He cells for a wide variety of applications. The majority of the cells have been fabricated from the boron-free aluminosilicate glass GE180[8]. Other constructional materials have included the aluminosilicate glass Corning 1720[9], ^{10}B-depleted Corning 1720, Corning 7056, fused quartz, pyrex, and silicon.

2.1. Preparation at glass shop

Most of our cells are blown by the NIST glassblower on a lathe, beginning from either 16 mm or 25 mm diameter GE180 tubing and then reshaping to a desired geometry, typically a cylinder with a volume up to 1 liter. The completed cells are immediately placed into an annealing furnace at 785 °C (GE180) for about 10 min to remove residual stress in the glass. The glassblower attaches one or two cells to a glass manifold referred to as the "string"(see Fig. 1), which also has one or two vertical tubes ("pull-offs") for alkali-metal ampoules at the end of the string. Between the cell and the pull-offs is a small dip in the string, referred to as the "retort", into which alkali-metal will be distilled before being distilled into the cell. The string and attached cell(s) are first checked with a polariscope to ensure sufficiently low residual stress. To minimize the possibility that the cell will rupture in the SEOP oven, the cells are then tested at room temperature to a pressure 25 % higher than the pressure expected under SEOP operating conditions.

2.2. Cleaning and first distillation

To make practically useful cells with long relaxation time (T_1) and desired performance, we have followed an extensive cleaning and baking procedure to ensure a cell has a minimal amount of impurities adsorbed to the surface of the glass, potentially yielding a long ^{3}He polarization relaxation time. Nevertheless, cell to cell variations and the exact significance (or lack thereof) of aspects of the procedure are not well understood. The principles are to clean the cells with simple soap and water, bake out impurities from the cell and string, and then do a double distillation of the alkali-metal. A clean location is prepared for the alkali by baking, the alkali is distilled into this location, then the cell and string are baked again before the alkali is distilled from this location into the cell.

All cells are thoroughly cleaned before they are attached to the vacuum system. The general cleaning procedure involves, in the following order: one rinse with soapy water (laboratory glassware detergent powder and water), several rinses with distilled water, one with acetone, several more rinses with distilled water, and then a final rinse with methanol (primarily to remove the water). More recently we have eliminated the acetone step, with no obvious ill effects. After the string is completely dry it is attached to an oil-free, high vacuum system (referred to as the cell filling station, see Fig. 2) consisting of a turbomolecular pump backed by a diaphragm pump. After leak-testing the string (often done quickly with alcohol), the pre-scored alkali metal ampoule(s) (Rb and/or K) are opened in a weak, outgoing flow of N_2, and dropped with

Journal of Physics: Conference Series **294** (2011) 012003 doi:10.1088/1742-6596/294/1/012003

open-end down into the pull-offs. The N_2 flow is then stopped and the top of the pull-offs are sealed off using a natural gas torch. At this point the cell(s) and string (except for the pull-offs that contain alkali-metal) are baked at 400°C for a total of four days to minimize impurities on the surface of the glass. When beginning baking, it is better to heat the cell and string gently to 400°C and ensure the temperature gradient near the cell is small. The pressure in the vacuum system will initially increase with the increasing temperature, then decrease slowly until approaching a vacuum of 10^{-7} mbar or better. After baking the string a day or so, the Rb or K is slowly melted out of each ampoule by gently heating the lower sides of the pull-offs. The pull-offs, which are not in the baking oven, are also gently heated with the torch until the pressure rise is minimal. After the alkali metals are melted out of the ampoules, the vacuum will reach about 6×10^{-8} mbar in another day of pumping.

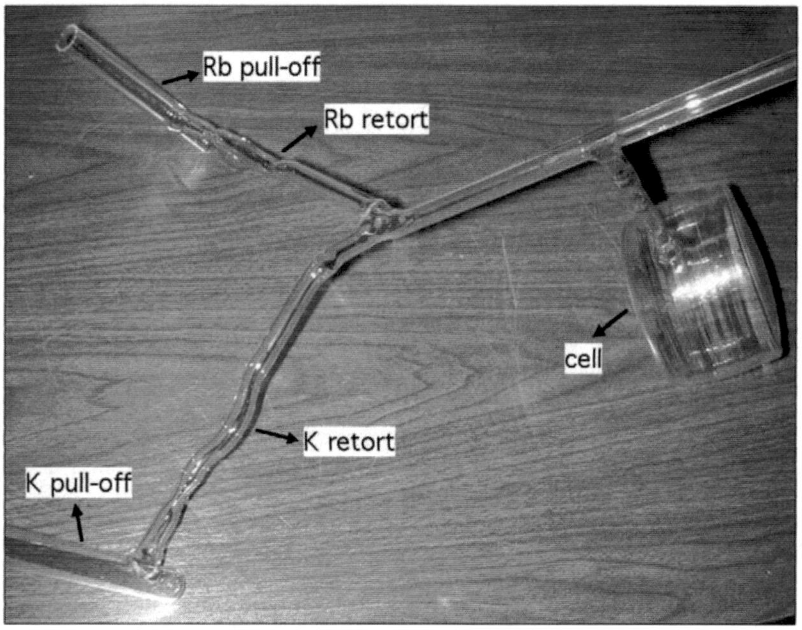

Figure 1. A glass manifold ("string") with cell attached. One or two (one for pure Rb cells two for Rb/K mixtures) vertical tubes ("pull-offs") for alkali-metal ampoules are located at the end of the string (left). On each arm of the string (two arms shown for this mixture string) there is a small dip, referred to as the "retort", into which alkali-metal will be distilled before being distilled into the cell. The string is attached to the vacuum system by a glass to metal seal followed by a metal-sealed flange (right, not shown).

On the third day the baking is stopped so that portions of the Rb and/or K can be "chased"(evaporated and re-condensed several times to move the metal along the string) from the pull-offs and into the retort. The pull-offs are then removed by sealing the string under vacuum, and baking is continued. For a pure alkali-metal (Rb or K) cell, we typically chase about 1/8 to 1/6 of the alkali-metal from the 1 g ampoule into the retort. For a Rb/K hybrid cell, we typically chase much more (about 1/2) of the K from its ampoule into the K retort. During chasing, it is better to distill the alkali metals in the vapor form many times. The cells are baked for two more days.

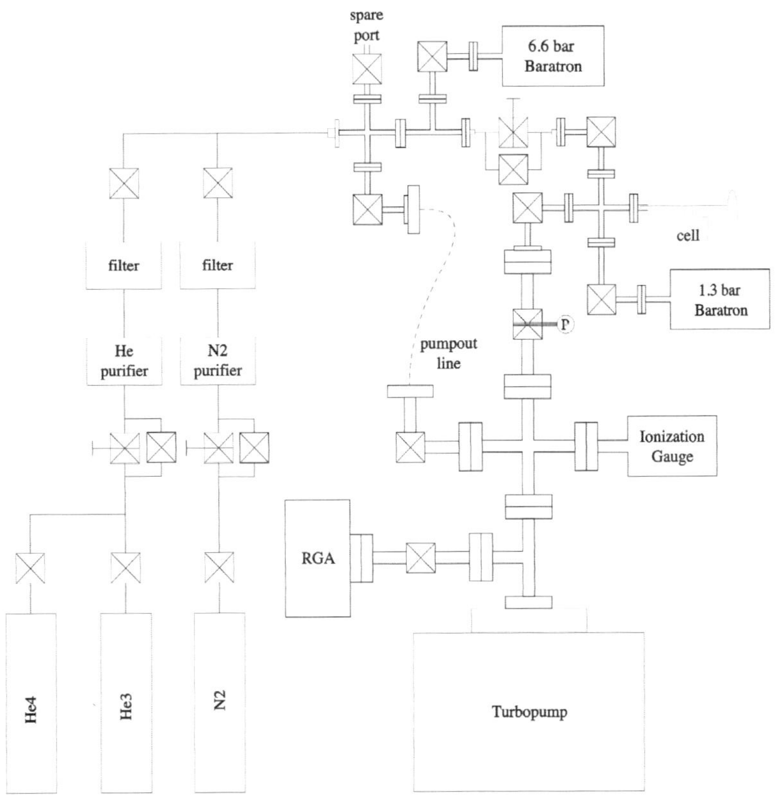

Figure 2. A schematic diagram for the cell filling system. It consists of three main sections, the oil-free high-vacuum system, the gas-handling system, and the cell-baking unit. The outside dimensions of the tubing in the diagram are 7 cm diameter (large), 3.4 cm diameter (small), and 6 mm (gas lines, shown as single lines); all connections are metal-sealed. The gas-handling system provides ^{3}He, ^{4}He, and N_2; each gas line has a getter and a particle filter for purifying the gas, and a needle valve (with bypass for pumpout) for flow control. It can be isolated from the other systems. The lower-conductance pump-out line is used for initial pump-out of a string and for pumping gases from the gas lines without affecting the string vacuum. The cell-baking unit (not shown) has two types of furnaces for different baking stages, a larger one for initial baking and a smaller one for final baking. It can accomodate three cells simultaneously, and can also accomodate valved cells.

2.3. Distilling Alkali metal into the cell

For a pure Rb or K cell, distilling either Rb or K into the cell is simple and straightforward as described previously[10]. As reported previously[11], K/Rb vapor density ratios (D) between 2 and 6 have yielded the best results for high polarizing rate without losing the maximum achievable polarization. Cells with vapor density ratios below this range yield lower gain in SEOP efficiency although are still usable. However, in a cell with too high a vapor density ratio the alkali-metal polarization does not reach 100 %, with the limit decreasing with increasing vapor density ratio [3, 11]. Hence it is essential to control the vapor density ratio to the desired range although it requires more complicated distillation of alkali metals into the cell. For D

between 2 and 6, about 12 to 36 times more K than Rb must be distilled into the cell. We have developed a convenient way to achieve the desired vapor density ratio and check it before the cell is filled with ^{3}He and N_2 gas and completely sealed off from the vacuum system. It was determined through trial and error that chasing Rb about six times in films fairly thick and about 5 cm long, then chasing about 1/5 of the K from the ampoule into the cell typically yields a vapor density ratio in the desired range between 2 and 6. After chasing, the Rb in the cell should have a slight translucent tint to it. It is better that after the Rb is distilled in the cell, the string for the K path to the cell is cleaned free of Rb, as the final vapor density ratio is sensitive to Rb that may be carried into the cell with the K.

It is crucial to measure and check the vapor density ratio on the vacuum system before the hybrid cells are removed from the string. We illuminate the cell with a 50 W halogen spot light and use an optical spectrometer with a wavelength range from 755 nm to 830 nm and a resolution of 0.07 nm[12] to measure the ratio of the absorption of K and Rb for each fine-structure transition (see Fig. 3). The cell is cycled to typical SEOP temperatures to establish the Rb/K alloy. In vacuum, the absorption will saturate in a bandwidth that is much smaller than either the resolution or 0.03 nm pixel separation of the spectrometer, hence the absorption is difficult to observe. Introducing gas broadens the response, yielding more easily visible absorption. We typically fill the cell with N_2 gas to a pressure of between 200 mbar and 260 mbar, then heat the cell to about 85 °C in the baking oven with a clear path for light absorption spectroscopy. There are four related fine-structure transitions in the spectrometer wavelength range, Rb $5s_{1/2}$-$5p_{1/2}$ (denoted as Rb D_1) and $5s_{1/2}$-$5p_{3/2}$ (denoted as Rb D_2), and K $4s_{1/2}$-$4p_{1/2}$ (denoted as K D_1) and $4s_{1/2}$-$4p_{3/2}$ (denoted as K D_2). Recently we have used a dual channel optical spectrometer[13] that allows us to observe both the $5s$-$5p$ and $5s$-$6p$ (D_2 at 420.18 nm and D_1 at 421.55 nm) transitions for Rb and $4s$-$4p$ and $4s$-$5p$ (D_2 at 404.41 nm and D_1 at 404.72 nm) for K. The temperature of the cell is adjusted so as to obtain absorption dips for these four transition of 10 % to 40 % (typically 85 °C for the infrared transitions or 170 °C for the blue transitions). Because mixture ratio determinations from the blue transitions are performed closer to actual optical pumping conditions, they require less correction for the different slopes of the vapor pressure curves. If the vapor density ratio is lower than 3 and higher than 6 after initial distillation of both Rb and K into the cell, it is necessary to either chase extra Rb into the cell for a vapor density ratio higher than 6 or bake the cell at some modest temperature for a vapor density ratio less than 3. The idea of baking is to remove more Rb out of the cell than K by fractional distillation, hence improving the vapor density ratio. The baking time should strongly depend on the temperature, varying from half an hour at 150 °C to overnight at 80 °C. We have successfully employed this approach, but have not experimented with the required baking time much yet. The vapor density ratio is re-checked until it falls within the desired range.

2.4. Filling the Cell with Gas

Before the cell is filled with gas, the gas lines are evacuated, and the ^{3}He and N_2 getters are heated to about 350°C. After about a hour, the vacuum is typically below 10^{-7} mbar. Both the nitrogen and ^{3}He gas lines have needle valves that are used to control the flow rate through the getters and a third needle valve controls the flow rate into the cell. Each needle valve has a bypass line which is closed except during pumping of the gas lines. We first add between 65 mbar and 130 mbar of nitrogen into the cell, and then close the valve between the cell and the gas lines, and pump out the nitrogen. To avoid a return flow of nitrogen into the ^{3}He lines, we admit ^{3}He gas until the pressure read on the 6.6 bar gauge is higher than the nitrogen pressure, and then open the valve to the cell to fill the appropriate amount of the ^{3}He gas into the cell. The

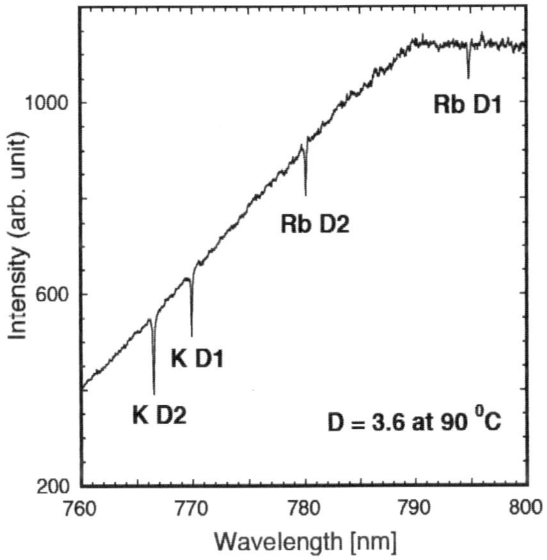

Figure 3. Light absorption spectroscopy for measurements of the vapor density ratio of K to Rb. The four absorption dips correspond to the four atomic transitions (Rb D_2 and D_1 and K D_2 and D_1) as discussed in the text. The K/Rb vapor density ratio D is determined from these four absorption dips. Shown here is the vapor density ratio measurement for the cell Syrah at 90 °C. The variation with wavelength of the transmitted light is primarily due to the spectrometer response.

amount of ^{3}He gas depends on the neutron wavelength, the length of the cell and the practical pressure limit for the cell. There are two approaches to filling the cell with ^{3}He, depending on how much gas in total (^{3}He, N_2, possibly ^{4}He) is required. If the desired total pressure is less than 1 bar, the gas filling procedure is straightforward; we simply admit nitrogen gas first and then ^{3}He gas into the cell. For cells with an overall pressure greater than 1 bar, it is required to completely or partially submerge the cell into liquid N_2 (hence a maximum pressure achievable in the cell of $\approx$3.7 bar) to allow the cell to be sealed off from the string under 1 bar. However, we have found that using complete immersion is not reliable for large cells because of the buoyant force and close proximity of the hot torch and liquid N_2. To make the seal more reliable for the typical final pressure range of 1 bar to 2 bar for neutron spin filters, we submerge only a small portion of the cell in the liquid nitrogen. We fill an appropriate total pressure, $p_{\rm i}$, of nitrogen and ^{3}He gas (N_2 first, then ^{3}He) into the cell at room temperature. With the cell and string isolated from the vacuum system, we gradually submerge the cell into the liquid nitrogen until a final pressure, $p_{\rm f}$, of about 0.9 bar is obtained. Typically, the cell is submerged no more than a couple of centimeters into the liquid nitrogen and thus the final seal from the string can be easily done. The total pressure of N_2 and ^{3}He gas in the cell, $p_{\rm c}$, after the cell is removed from the string and equilibrated to room temperature is given by

$$p_{\rm c} = p_{\rm i} + \frac{V_{\rm s}}{V_{\rm c}}(p_{\rm i} - p_{\rm f}) \tag{1}$$

where $V_{\rm c}$ and $V_{\rm s}$ are the volumes of the cell and the string, respectively. For a large number of cells, we have found that the values of $p_{\rm c}$ determined with this method agreed with those determined from neutron transmission[14] within 5 % (see Fig. 4). Data points in Fig. 4 are scattered around the line that represents the same pressure from both methods and has a slope of unity.

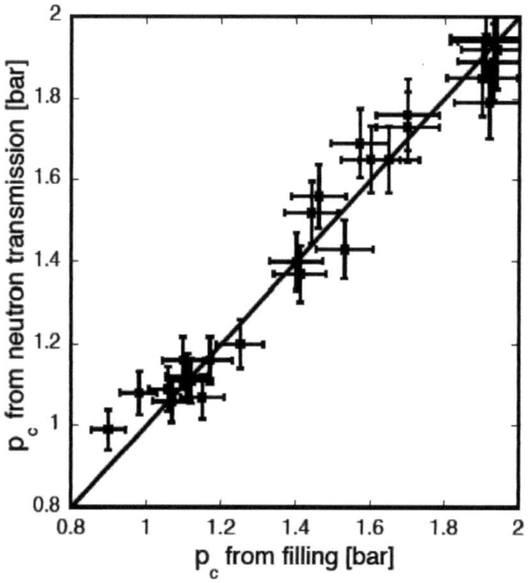

Figure 4. ^{3}He partial pressure in the cell obtained from both filling and neutron transmission measurements. The solid line represents an equal ^{3}He pressure from filling and neutron measurements. We estimate the relative standard uncertainty for both ^{3}He pressure determinations from cell filling and neutron transmission measurements to be 5 %. For transmission-based values the uncertainty is dominated by the cell length determination.

3. ^{3}He cell performance tests and diagnostics

After the cell is removed from the string, there are extensive tests and measurements to perform, including the pressure test at high temperature; the vapor density ratio check; ^{3}He polarization relaxation time, T_1, and its orientation dependence [4] if applicable; and maximum achievable ^{3}He polarization and related value of X. (X is a parameter that describes the recently observed linear increase in relaxation with alkali density, which limits the maximum ^{3}He polarization[15].) This section simply describes how these tests were done. The pressure test at high temperature is an additional safety check to minimize the possibility of cell rupture in the SEOP system. The hot pressure test is carried out in a well-controlled environment at a temperature 20 °C higher than the maximum temperature expected for optical pumping. After the cell passes the hot pressure test, we check the vapor density ratio on one of the SEOP systems using the protocol mentioned earlier. We found the initial vapor density ratio measurement for the sealed cells agreed with the measurement done on the cell filling system within 15 %. However, the vapor density ratio can change (typically an increase) during the first few cycles of cell heating to optical pumping temperatures presumably because the Rb-K alloy is not fully mixed.

3.1. SEOP systems

We currently have four SEOP systems at NIST used for neutron science applications. The first one[16, 11] has been used for over a decade for research and development. During this time period we have transitioned from using a single, 30 W, fiber-coupled, broadband laser to using two or three spectrally narrowed lasers, each based on a 50 W or 100 W diode array bar. It is currently employed primarily for the development of wide-angle analysis[7], but is also being used for other activities such as tests of large double-chambered cells. Two newer systems, based on the first system, are dedicated to providing polarized ^{3}He gas for polarized neutron experiments at the NIST Center for Neutron Research (NCNR)[17]. Finally we have a fourth compact system on the NG6M beam line[14] that was recently employed for polarizing small cells for neutron interferometry[18, 19]. The first three systems employ 795 nm spectrally narrowed diode lasers to illuminate cells from opposite sides. Use of a polarizing beam splitter just before the quarter-wave plate cleans up any loss of ideal linear polarization due to optical elements and also rejects light originating from the laser on the other side of the cell. In the first system we currently have three lasers in use. Two employ water-cooled single bar diode arrays, each narrowed with an external cavity. The first is based on a 50 W bar[20] that operates at minimum feedback, whereas the second is based on a 100 W bar[20] and employs a polarized beam splitter cube in the external cavity[21]. In the 100 W system we employ a half-wave plate between the two telescope lenses to control the feedback to the laser. The optic axis of the half-wave plate is typically adjusted to rotate the linear polarization of the laser light by 25°-30° relative to the zero-feedback condition. The third is a commercial fiber-coupled laser narrowed by a volume Bragg grating (VBG) and temperature-controlled with a thermoelectric cooler[22]. This system is convenient and compact, but for our particular unit we have had trouble obtaining ideal overlap with the proper wavelength. In each of the two NCNR systems, we employ two 100 W bars, each with an external cavity with an internal polarizing beamsplitter. Both bars are water-cooled by a single chiller[23], with flow control to allow operation at different temperatures. The lasers are interlocked such that loss of water flow will shut down the lasers. All the key components in these systems including Helmholtz coils; laser, optics and chiller; heating unit; and polarization monitoring apparatus (see below) are integrated into a portable aluminum[24] cart. The laser shaping optics in these SEOP systems[16] can be conveniently adjusted to fully illuminate cells with cross sections up to 12 cm by 12 cm. We have recently replaced the external cavity in one laser system with a chirped VBG[25]. A chirped grating can be tuned about 1.5 nm, which avoids the wavelength matching issue sometimes encountered with a fixed VBG. We have obtained a linewidth of 0.25 nm full width at half maximum (FWHM) with excellent suppression of broadband background and only $\approx$10 % loss of the free-running power output[26]. Due to heating of the grating, the central wavelength should be specified below the desired transition. For 50 W and 100 W bars we have employed a grating with a central wavelength of 794.7 nm (vacuum wavelength; Rb transition at 795.0 nm). Although this specification was close to ideal for the 50 W bar, the laser beam is near the grating edge for the 100 W bar. We have recently tested gratings with central wavelengths of 794.1 nm (vacuum), and found that this wavelength specification is preferable for 100 W bars. We expect to convert more of our off-line lasers to operation with chirped gratings and modify the beam shaping optics to optimize the performance in this configuration. Fiber-coupled, VBG-narrowed lasers with thermoelectric cooling promise convenience for on-line operation.

3.2. Polarization measurement and monitoring

The ^{3}He polarization during optical pumping or relaxation is monitored by either the adiabatic fast passage (AFP) or the free induction decay (FID) nuclear magnetic resonance (NMR) methods [27, 28]. The AFP NMR data is obtained by sweeping the magnetic field that maintains the ^{3}He polarization through the ^{3}He resonance frequency (typically 70 kHz to 90 kHz) and sweeping back to restore the original ^{3}He polarization state. The AFP data is fit to the square root of a Lorentzian, with a background term that includes a possible linear drift. This yields an NMR amplitude proportional to the ^{3}He magnetization. To obtain the absolute ^{3}He polarization, the AFP NMR is calibrated for each cell against neutron transmission measurements[14]. We have also compared the neutron-based calibration to that obtained from an absolute calculation of the AFP signal based on the known geometry and measurements of the pick-up coil quality factor. For small cells we find that the AFP signal is typically 20 % lower than one would expect from the absolute calculation, whereas for large cells this difference can be between 20 % and 50 %. We often employ FID NMR during optical pumping or relaxation because it does not require nulling of the signal induced in the pick-up coil and thus is more easily automated. Fig. 5 shows typical NMR signals and pumping and relaxation curves. Both AFP and FID for both SEOP systems are controlled by a single NMR unit, which may be expanded in the future.

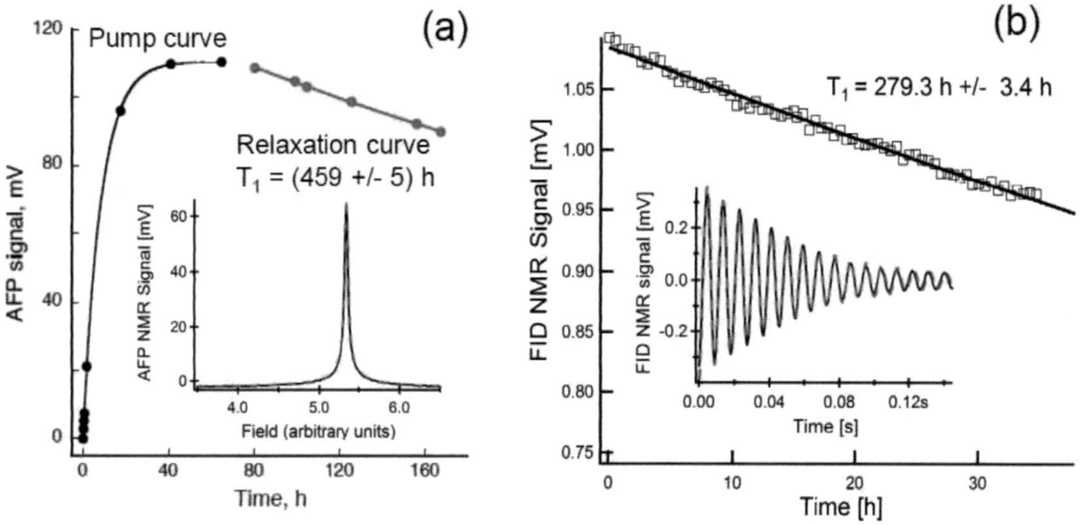

Figure 5. Typical NMR signals and pumping and relaxation curves. (a) Pumping and relaxation curves determined from AFP NMR; inset shows an AFP signal. AFP signals are calibrated against neutron transmission measurements to yield absolute ^{3}He polarization values with typical relative standard uncertainties of 4 %. (b) Relaxation curve from FID signals shown in the inset.

4. Discussion of cells

4.1. Survey of relaxation characteristics

In the appendix we provide a detailed listing of NIST cells and applications. Here we survey the wide range of behavior we have observed for relaxation times. Fig. 6 shows the room temperature T_1 of the blown GE180 cells for the ^{3}He partial pressure pressure range between 0.13 bar to ≈2 bar. Several cells have relaxation times approaching the dipole-dipole limit[29], with the longest T_1 of 3000 h obtained in the cell Diamond[30]. For the ^{3}He partial pressure of 0.13 bar for this cell, this result indicates the wall relaxation contribution to the total relaxation is $\approx$ 6000 h. The cells Diamond and Emerald had ^{4}He added to reduce relaxation from magnetic field gradients[31].

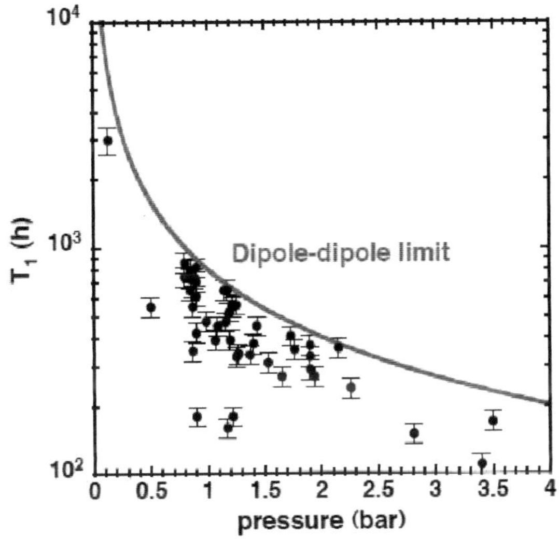

Figure 6. The room temperature relaxation time vs the partial ^{3}He pressure for the blown GE180 cells made at NIST. The solid line shows the limit from dipole-dipole relaxation. The longest relaxation time was 3000 h for the cell Diamond with a partial ^{3}He pressure of 0.13 bar. The error bars show the typical relative standard uncertainty of 10 % or less in fitted values of T_1.

Many cells, in particular those for which the relaxation is dominated by dipole-dipole relaxation, exhibit a stable relaxation time. However, many cells often show low field orientation dependence[4], which may or may not converge to a single, stable value after repeated cycles of optical pumping. More generally, a cell that exhibits a dependence of its relaxation time on its orientation with respect to a magnetic field may also show a dependence of its relaxation time on the magnitude of the magnetic field, even for fairly weak magnetic fields. Jacob[32] observed minima in the relaxation time of a GE180 SEOP cell at coercive fields of $\approx \pm 200$ G. In Fig. 7 we show magnetic field effects on relaxation for two cells. In Fig. 7(a) the wall relaxation rate for the cell Spock shows a variation between 0.004 h^{-1} and 0.01 h^{-1} for fields between -2.6 mT and 2.6 mT, with the longest relaxation time recorded at 1.15 mT. For these studies an FID signal was measured at 2.6 mT with the bipolar power supply for the Helmholtz coils operating in current control mode, then the supply was switched to voltage control mode (set to maintain nearly the same current) and the voltage was varied to produce fields between -2.6 mT and 2.6 mT for some period of time. Finally the voltage was returned to produce 2.6 mT and the supply was switched back to current control mode for another FID measurement. We tested that quickly passing through zero field and switching modes of the power supply did not affect the ^{3}He polarization.

To investigate the effect of stronger fields we used a makeshift set of four small coils to produce the results in Fig. 7(b) for the cell Skylight. This cell had a reasonably long relaxation time of T_1=156 h, as measured in the Helmholtz coils of our SEOP apparatus operating at $\approx$2.5 mT. Exposing the cell to a field of 43 mT and returning to check the T_1 at 2.5 mT decreased

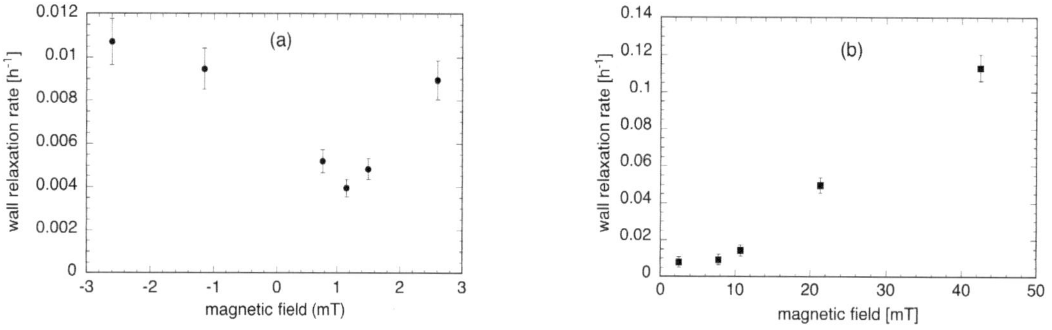

Figure 7. Dependence of the relaxation rate on magnetic field for two cells. (a) Wall relaxation rate for the cell Spock for fields between −2.6 mT and 2.6 mT. The data shown were corrected from the measurements by the dipole-dipole relaxation rate of 0.0011 h^{-1} for ^{3}He partial pressure of 0.9 bar in this cell. (b) Wall relaxation rate for the cell Skylight for magnetic fields between 2.5 mT and 43 mT. The data shown were corrected for dipole-dipole relaxation (0.0037 h^{-1} at 3.0 bar) and gradient-induced relaxation (0.013 h^{-1} for the relatively large value of 2×10^{-3} cm^{-1} in the makeshift four-coil apparatus employed). For both plots the error bars indicate the estimated uncertainty in the measurements due to stability of the NMR signals, and for (b) also include uncertainty in the correction for gradient-induced relaxation.

the T_1 to 100 h, and the subsequent series of measurements obtained at fields from 2.5 mT up to 43 mT showed over an order of magnitude decrease in relaxation time with increasing field. For these tests the AFP signal from the cell was measured in our Helmholtz coils, then the cell was transferred directly to the nearby four coil system operating at values between 2.5 mT and 43 mT and maintained for some period of time, and finally the cell was transferred back to the Helmholtz coils for another NMR measurement. While the cell was in the four coil system the Helmholtz coils were turned off to avoid affecting the measured T_1, and while the cell was in the Helmholtz coils the four-coil system was turned off to avoid affecting the AFP signal. Whereas such effects will surely be cell-dependent, these effects must be considered for any application in which one expects cells to experience magnetic fields greater than ≈10 mT.

Long relaxation times are most reliably obtained for fully blown cells, but can be obtained in flat-windowed cells. Cells with short relaxation times can often be improved by opening the cell, reacting away the alkali metal, and completely repeating the cell filling process. For some of the cells listed below, this procedure was performed one or more times. In some cases the cell is rinsed or soaked in nitric acid after the alkali-metal is removed. For one cell (Pike) our glassblower "flame-polished"the inside of the windows, and then an optical seal in the middle of the cylindrical section was used to seal the cell. The goal of this construction was to see if flame-polishing improved the relaxation time; the results were inconclusive and due to the complex constructional procedure we did not pursue this further.

In the case of the small flat-windowed cells Cashew, Peanut, Almond and Pistachio that were fabricated for neutron interferometry experiments, the least understood behavior was observed, in particular for the cell Cashew. For Cashew and Peanut we observed relaxation times of 350 h and 53 h, respectively, whereas the relaxation times for Almond and Pistachio were very short (≈1 h). These cells were not used for two years until the interferometer experiment was ready for them. Upon testing them, we found that the relaxation times for Cashew and Peanut had decreased to between 1 h and 5 h. We attempted to demagnetize the cells using a small demagnetizing unit typically employed for demagnetizing tools. In addition, in desperation, we made various attempts to redistribute the rubidium, and repeatedly optically pumped the cell. We were able to increase the relaxation times of these two cells to about 8 h. Finally Cashew was optically pumped "upside-down", in an attempt to cause greater flow of rubidium over the cell walls. The relaxation time of the cell increased dramatically to the listed value of 330 h, similar to its original value, and remained at this value during the course of use for the experiment. We cannot say what caused the change, but speculate that there is a highly relaxing spot on the cell that was eventually covered with alkali-metal. In the case of Peanut we were not able to exceed 8 h. Due to this strange behavior, we demagnetized and tested the originally bad cells Almond and Pistachio, and obtained relaxation times of about 35 h for both. The cells Cashew and Pistachio were employed for the interferometer experiment, with most of the data obtained using Cashew.

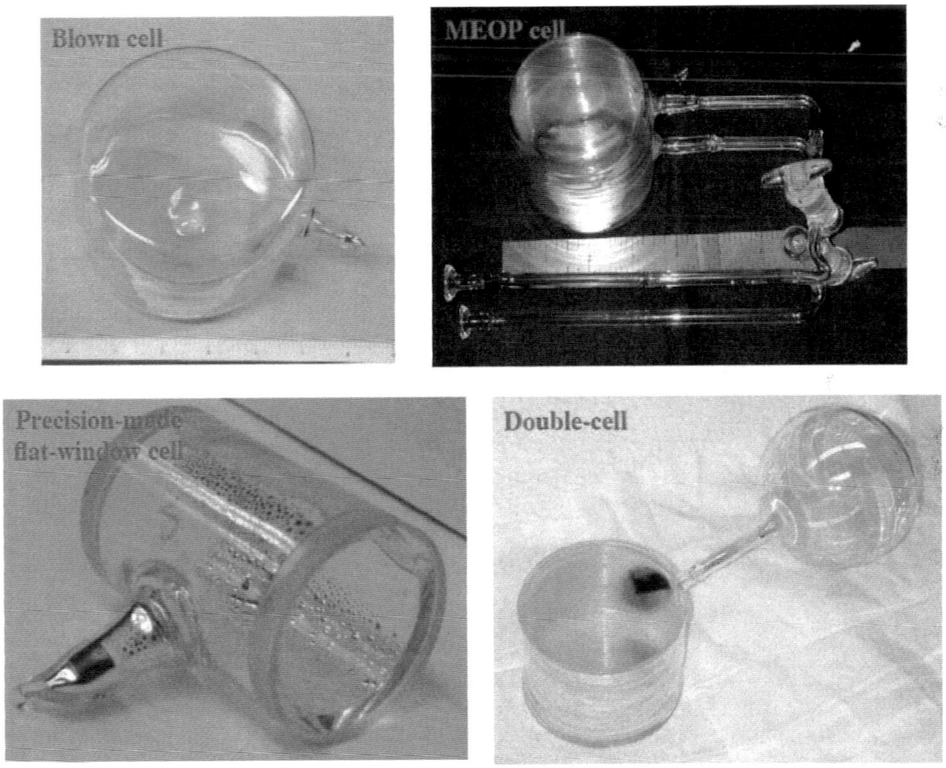

Figure 8. Photographs of a typical large blown cell (Barbera, 12 cm diameter), a valved MEOP cell (Orion, 7.5 cm diameter by 10 cm long), flat-windowed interferometer cell (Cashew, 2.3 cm diameter by 3.4 cm long) , and a large double cell (Sarek, 11 cm diameter spherical optical pumping chamber and 10 cm diameter flat-windowed spin filter cell). See the text and appendix for additional information on these cells.

MEOP cells provided additional information on cell relaxation times. For example, the Rb in the cell Arcturus became substantially oxidized due to slow leakage of air when on the shelf for a long time without exercise of the stopcock valves. Nevertheless, the relaxation time was measured to be 130 h, similar to the value of 170 h measured when the cell was first constructed. This result, along with other similar experiences we had with SEOP cells that developed accidental leakage, indicate that Rb oxides provide a low relaxation surface. This is expected as studies have shown excellent relaxation properties for cesium oxides[33]. In Ref. [10] we reported a relaxation time of 74 h for the cell Spica, whereas a relaxation time of 240 h was reported in later work[4]. This may have been due to slight oxidation that occured while the cell was on the shelf for a year or more. As described in Ref. [4], the cell Spica was then employed to establish the role of heating in low field orientation dependence. Before heating this cell showed an average relaxation time of 240 h with little orientation dependence, whereas after heating the relaxation time varied with orientation between 119 h and 258 h.

4.2. Special cell constructions

Fig. 8 shows photographs of a few cells: a typical, blown cell for neutron spin filter applications (Barbera); a flat-windowed cell for interferometry (Cashew), a valved, alkali-coated, MEOP cell (Orion), and a large double cell (Sarek).

A few other special types of cells are worth noting. In principle, Rb/Na hybrid cells should yield even greater efficiency than Rb/K cells[34]. However, due to the low vapor pressure of Na, extremely low mole fractions of Rb and operation at temperatures over 300 °C is required. We filled three Rb/Na cells, two from GE180 (denoted Margarita and Merlot) and one from Corning 1720 (denoted Shiraz). Our first cell, Margarita, contained too much Rb and thus was essentially a pure Rb cell with a Na coating. The value of 76 % polarization obtained in Margarita indicates that X is comparable to that obtained in the absence of the Na coating. The cells Merlot and Shiraz (not listed in the appendix) had more appropriate mixture ratios, allowing some studies before these cells developed an opaque, brown film. Sapphire can accomodate these temperatures, and a 24 h relaxation time has been reported for a pure Rb cell made completely from sapphire[35]. We filled two Rb/Na cells that employed sapphire end windows bonded to glass[36], but these yielded unacceptably short relaxation times of less than 10 min[37]. We have not yet attempted to refill these cells or construct others.

Silicon is an ideal material for neutron spin filter windows because of its high neutron transmission. For MEOP, relaxation times $\geq$240 h have been obtained in cesium coated cells constructed from silicon windows bonded to Pyrex[38]. We have tested two SEOP cells (denoted by Horta and Lazarus), that were constructed by the NIST Optical Shop from silicon windows and blown Pyrex. The relaxation times of these cells are in the 50 h range, but in some cases have shown substantial orientation dependence. We have not listed polarization values because of limited studies of these cells. However, the maximum values of $\approx$50 % polarization that we have observed to date suggests that they may show higher than average relaxation at high temperatures. Recent studies for SEOP Pyrex cells[39] indicate that they may exhibit higher than average X values, in particular for pure Rb cells. Further studies of the temperature dependence for Pyrex cells are needed. However, for larger silicon-windowed cells X may be sufficiently low so that polarization values comparable to those obtained in GE180 cells can be achieved. The cell Lazarus was reconstructed as a double cell so that the silicon-windowed part would not have to be heated. Nevertheless, this cell still showed orientation dependence, which we plan to study further in the future.

The recent discovery of neutron beam induced alkali spin relaxation[40, 41, 42, 43] presents a new issue for the use of spin filters in high flux beamss. Whereas not a major concern for most neutron scattering applications, this phenomenon must be considered for many fundamental neutron physics experiments. Double cells present a method for decoupling the optical pumping

Journal of Physics: Conference Series **294** (2011) 012003 doi:10.1088/1742-6596/294/1/012003

from the neutron beam, as has been done for targets for electron beams[44, 45]. Successful results were obtained in neutron tests at the Institut Laue Langevin using the double cell Roadrunner[42]. For further work in this area we have constructed a large double cell, denoted by the name of the original single cell that is used for the spin filter section, Sarek. This cell has a $\approx$700 cm^3 spherical optical pumping chamber that is connected to a 530 cm^3 flat-windowed filter cell through a tube that provides a 0.7 h transfer time for diffusion of the gas between the cells. Due to the 30 cm overall length of this extended cell, we employed an electrically heated oven located within a 40 cm diameter, 70 cm solenoid (prototype for the MACS apparatus) for tests. Despite a relaxation time of 150 h in the optically pumped orientation, we have observed only 46 % polarization in this cell. This result is lower than expectations, and is being studied further.

5. Cells for wide-angle polarization analysis

Wide-angle neutron polarization analysis presents new challenges for cell fabrication. Substantial progress has already been made for MEOP-based wide-angle spin filters[46], but directly optically pumped SEOP wide-angle cells present additional issues due to the need for operation at elevated temperatures. Although ideally we would fabricate such cells from GE180 glass, to date this has not been possible due to the difficulty in working this glass and the limited forms in which it is available. Before reporting our results for fused quartz wide-angle cells, we review some basic information about glasses for polarized ^{3}He.

In early cell development, aluminosilicate glasses were found to be preferred for storage of polarized gas in bare glass cells because of the low ^{3}He permeability for such glass. Fused quartz and borosilicate glasses were found to exhibit relaxation related to their higher permeability[47]. For fused quartz the permeability is sufficiently high such that ^{3}He loss for extended operation at SEOP temperatures is an issue. For these reasons and also because of good alkali resistance, aluminosilicate glass has also been used for SEOP. It has been shown that the alkali metal in SEOP cells improves relaxation time[48], which has led to the routine use of cesium as a coating for MEOP spin filter cells. Once alkali is introduced, high permeability glasses can be tolerated, indeed very long relaxation times have been obtained in cesium coated cells fabricated from fused quartz[49]. For magnetic resonance imaging applications, long relaxation times have also been obtained in valved Pyrex cells[50]. Nevertheless, the longest relaxation times are still most reliably obtained in aluminosilicate glass cells[10]. To explain this observation, it has been speculated that the wall relaxation observed in alkali-coated cells is due to imperfections in the coating. Such imperfections would be expected to have a more detrimental effect on the relaxation time for a high-permeability glass[33]. Such considerations become particularly critical for SEOP, as permeability increases exponentially with temperature. Indeed it has recently been shown that fused quartz cells show significant temperature-dependent ^{3}He relaxation that varies from cell to cell[5].

We have constructed two types of fused quartz cells to yield 120° of angular coverage[7]: three commercially made[51], welded cells (7 cm $\times$ 7 cm inside dimensions for cross section), and two cells made at the NIST shop from slightly resized, 7 cm or 8 cm inner diameter GE124 quartz tubing[52]. One of each type, denoted by the names Rainbow (welded) and Jersey is shown in Fig. 9. We employed pure Rb for the fused quartz analyzer cells due to browning of the glass in the presence of Rb-K vapor at temperatures above 200° C. (The brown coating rinsed off when the cell was opened and cleaned.)

For all three welded cells (denoted by Rainbow, Crescent and California), we observed significant, persistent orientation dependence in the relaxation time[4]. For example, for the wide angle cell Rainbow, we repeatedly obtained a relaxation time of 20 h when it was cooled to

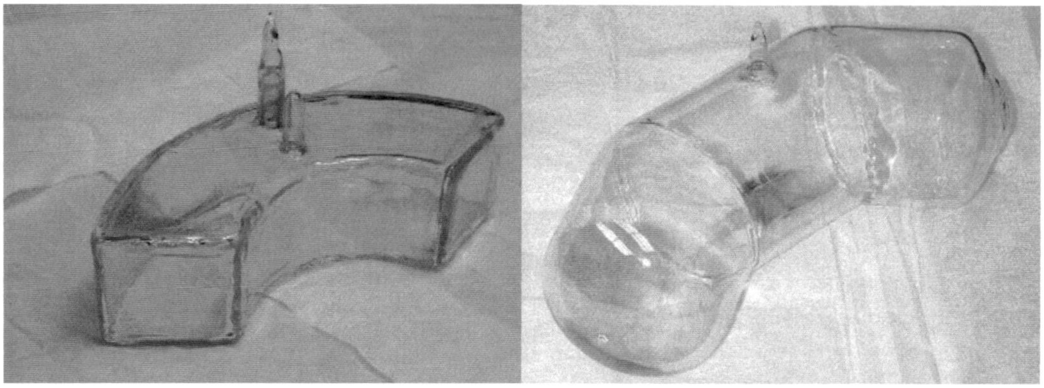

Figure 9. One of each type of fused quartz cells for wide-angle analysis cells: welded cell Rainbow (left) and cell Jersey (right) made from three connected sections of slightly resized tubing.

room temperature after optical pumping, and 90 h when the cell was then rotated 180°. When this cell was optically pumped at 170 °C, we obtained an optical pumping time constant of (8.4 ± 0.1) h. Assuming a typical value of $X = 0.3$[15] for this size cell, a maximum polarization of 0.45 was expected, in reasonable agreement with the measured polarization of 0.42. The polarization values were determined with a typical relative standard uncertainty of 5 % from neutron transmission measurements and/or electron paramagnetic resonance [53, 14].

It has generally been found that fully blown GE180 glass cells typically have smaller wall relaxation rates than non-blown cells[29, 10]. Hence to improve the relaxation times in our wide-angle cells we made three additional cells, denoted by Jekyll, Peru and Jersey, by fusing together 2 to 4 sections of fused quartz tubing. Each section was slightly resized so as to yield blown surfaces. This fabrication approach yielded a cylindrical test cell Jekyll (7 cm inner diameter (ID) by 13 cm long) and a wide-angle cell Jersey (8 cm ID cross section) with relaxation times of 210 h and 180 h, respectively. Whereas we would normally obtain over 70% polarization for such long relaxation times, we obtained an unexpectedly low polarizations of 0.47 and 0.42 for Jekyll and Jersey, respectively. In addition, the pumping time constant was shorter than expected. For example, for Jersey, only (7.1 ± 0.2) h was measured, compared with an expected value of 11.5 h (assuming $X \approx 0.3$) for the optical pumping temperature of 170 C. We believe this result is due to the same decrease in relaxation time with temperature as was recently reported for fused quartz cells[5]. In SEOP cells, the relaxation rate has been found to increase linearly with alkali-metal density[15]. However, the data in Ref. [5] indicates that for fused quartz the measured relaxation rates increase substantially in the absence of any significant Rb density as the temperature is increased. We plan further T_1 tests in this temperature regime to test our hypothesis. This temperature dependent phenomenon is not presently understood. It may be related to the high permeability of the fused quartz. Although it appears that this temperature dependence may not exist in our welded cells, this is difficult to ascertain because of the short relaxation times observed in the optically pumped orientation.

To temporarily bypass this fused quartz issue we have constructed two blown, cylindrical, GE180 analyzer cells (Barolo and Barbaresco), which we are currently employing for the first experiments on MACS, albeit with reduced (30 °) angular coverage. As discussed in Ref. [7] the cells Monastrell, Gamay and Grenache will be used to polarize the MACS beam. On the longer term, we will make further attempts to fabricate wide-angle cells from GE180 and investigate coatings such as sol-gel to address the high temperature relaxation issue in fused quartz. Other alternatives include valved cells or double cells made from fused quartz, to avoid heating of the spin filter cells.

6. Conclusions

We have described our procedures for fabricating and testing SEOP cells, and reviewed and documented results for over a decade of development and application of polarized ^{3}He cells. The relaxation times for such cells show remarkably complex behavior that is not well understood. Each new goal in cell development or more careful study of existing cells reveals new phenomenon. While interesting, these phenomena can often be an impediment to applications, but somehow we always muddle through the maze. We presented the current state of the most recent challenge of wide-angle neutron polarization analysis.

Acknowledgments

We thank the NIST glassblower, Jeff Anderson, and Jack Fuller of the NIST optical shop, without whom this work would not have been possible. We also acknowledge G.L. Greene, M.S. Dewey, F.E. Wietfeldt, A.K. Thompson, and T.B. Smith for their contributions to the apparatus for preparing and characterizing cells, and V. Pomeroy for suggesting the use of GE180. In addition, we acknowledge E. Babcock, B. Chann, T.G. Walker, R. Jacob and B. Saam for collaboration on SEOP. The development and application of neutron spin filters has been supported during the time period of this work in part by the U.S. Dept. of Energy, Basic Energy Sciences.

Appendix

In this section we present a series of tables of NIST cells, divided into four categories: SEOP cells for tests and studies, MEOP cells, cells for neutron scattering applications at the NCNR, and cells for neutron scattering applications at other laboratories and fundamental neutron physics. Although this list is close to complete for cells made in the last decade, it does not include many older cells. There is of course overlap in these categories, as many cells have had multiple uses. For each cell, we provide references for where further information on the cells and/or applications can be found. Whereas NIST was involved in most of these applications, we also include applications of loaned cells.

Table 1 lists cells used for tests and studies, including development of long relaxation time neutron spin filters[10], SEOP with spectrally narrowed diode lasers[16], low-field orientation dependence of ^{3}He relaxation time in SEOP cells[4], limits to the polarization for SEOP[54, 15], hybrid SEOP[11], neutron beam effects on SEOP[40, 41, 42, 43], and evaluation of magnetic field gradients[55, 56].

Table 2 list storage cells used for metastability-exchange optical pumping, including development of compression systems[57, 58]; application to small angle neutron scattering (SANS)[59], reflectometry[60], and magnetic resonance imaging[61]; and studies of low field orientation dependence of ^{3}He relaxation time[4].

Table 3 lists cells employed for neutron scattering applications at the NIST Center for Neutron Research (NCNR)[14, 17], divided into three categories: 1) small angle neutron scattering

(SANS)[59, 62, 63, 64, 65, 66] and the NCNR reflectometers (REFL)[67, 68, 69], 2) triple-axis spectrometry (TAS)[70, 71, 72, 73, 74] and 3) the NCNR Multi-Axis Crystal Spectrometer (MACS)[7].

Table 4 lists cells used for neutron scattering applications on the Single Crystal Diffractometer (SCD) at the Intense Pulsed Neutron Source (IPNS) at Argonne National Laboratory (ANL)[75, 76], the magnetism reflectometer at the Spallation Neutron Source (SNS-MR)[77, 78], and the CRISP instrument at ISIS[21], followed by fundamental neutron physics applications[30]. The latter are divided into three categories: neutron interferometry at the NCNR[18, 19], the NPDGamma experiment at the Los Alamos Neutron Scattering Center (LANSCE)[79, 80, 81, 82, 83], and limits on short range spin-dependent forces from spin relaxation of polarized ^{3}He[84]. A detailed paper on the neutron interferometry experiment, including the precision polarimetry for this work[85], will appear in a future publication.

For all four tables the following information applies: D is the vapor density ratio for Rb/K cells (D=0 indicates pure Rb and D=∞ indicates pure K), d is the inside diameter in cm, and l is the average inside length in cm. Most cells are cylindrical, but the absence of a value for l indicates that the cell is spherical. An additional value in parentheses indicates the diameter of the spherical optical pumping chamber for double cells. V is the volume in cm^3, p is the ^{3}He partial pressure in bar, T_1 is the ^{3}He relaxation time in hours, and P_{He} is the maximum ^{3}He polarization observed for 795 nm optical pumping. For T_1, two values are listed where orientation dependence was measured at least once or a range of values is listed for more extensive studies of orientation dependence. For P_{He}, a number is parentheses for mixture or pure K cells indicates results for optical pumping with a 770 nm laser[11] based on either a 20 W or 40 W diode bar. The last column includes notes and references for further information and/or applications for each cell. If the P_{He} was never measured accurately (or if old records are inconclusive), there is no entry in this column. In addition, we note that not all cells were optically pumped under ideal circumstances. Relative standard uncertainties for data in these tables varied with the amount of studies of a given cell, but are typically equal to that indicated by the last significant figures or 5 %, whichever is greater.

In many cases where details on how the cells were pumped is available in the listed references. Except where noted all cells were constructed from GE180 glass; in the last column FQ indicates fused quartz, 1720 indicates Corning 1720[9], D1720 refers to a special batch of glass that was made with ^{10}B-depleted Corning 1720 glass, and 7056 indicates Corning 7056[9]. In several cases, the material for the body and windows for a cell may differ. Cells may be listed twice if they have been reconstructed at some point. The cells Lazarus and Sarek were originally fabricated as single cells, but later became part of a double cell with the name unchanged. The cell Mars was first fabricated as an valved, uncoated cell for MEOP and then later became a sealed, pure Rb SEOP cell.

Journal of Physics: Conference Series **294** (2011) 012003 doi:10.1088/1742-6596/294/1/012003

Table 1. ^{3}He cells used for tests and development. The cells are divided into two categories, those that were fabricated from blown glass and those that have flat, optically sealed windows[86].

Cell name	D	d	l	V	p	T_1	P_{He}	References, notes
BLOWN								
Bonnie	0	2.8	-	12	3.4	110		[15]
Ringo	0	2.5	-	8	3.4	64	0.48	[15]
Fred	0	4	4	50	0.8	≈170	35	[87]
Mrs. Peel	0	6	7	200	1.60	180	23	[87]
Barney	0	8.5	4	230	1.1	380		[10, 15]
Natasha	0	4.5	4.5	72	0.7	440		[10, 15], FQ
Red Baron	0	5.8	1.5	40	0.9	82/270	0.53	[4, 15]
Riesling	4.8	9.8	9.0	680	3.5	170	0.77 (0.68)	[11]
Cabernet	46	11.8	7.6	830	1.94	270	0.58 (0.71)	[11]
Gigantor	96	9.1	9.4	611	0.85	800	0.56 (0.73)	[11]
Quasimoto	155	6.7	5.8	205	1.12	550	0.48 (0.76)	[15, 11]
Sonora	∞	4.5	4.3	68	0.81	500	(0.77)	[11]
Otto	0	1.4	9	14.7	0.9	420	0.45	[15]
Uhura	0	5.0	5.4	104	1.22	180	0.64	[15]
Wilma	0	4.9	4.9	91	0.8	830	0.76	[10, 16, 15, 11]
Scotty	0	5.1	5.3	108	0.90	610		
Tin Man	0	4.9	4.9	91	0.58	70	0.77	[43]
Lion	0	4.7	5.0	86	0.57	510	0.77	[43, 88]
Scarecrow	0	5	5	90	0.5	25	0.22	[43], low N_2 pressure
Diamond	0	3	-	≈14	0.13	3000		[30], 0.77 bar ^{4}He
Emerald	0	2.5		≈7	0.13	≈2000		[56], 0.70 bar ^{4}He
Sunshine	0	11.2	-	740	0.5	550		[15]
Washington	0	2.2	4.8	27	0.87	550	0.65	
Margarita	≈0	6.4		140	0.9	820	0.76	[15], Na coating
Merlot	-	6.4		140	0.9	250		Rb/Na
FLAT								
Betty	0	4.7	4.9	81	0.87	240	0.72	[10, 54, 16, 4, 15], 1720 body
Sulu	0	4.8	6.2	110	0.9	115	0.63	[16, 15]
Pike	0	5.3	6.1	135	0.77	40/310	0.67	see text
Boris	0	4.7	4.9	84	0.85	85		[10, 15], 1720
Mars	0	6.0	8.0	230	0.9	270	0.73	[16, 4, 15], 1720 body
Spock	0	10.3	7.4	620	0.92	35-229	0.65	[4, 30]
Sarek	0	10	6.7	530	1.06	39-135		
BamBam	0	9.4	4.7	330	0.85	95	0.51	[16, 4, 15]
Algonquin	0	3.7	7.4	80	1.58	13-70	0.24	[15]
Horta	0	4.2	5.6	81	1.3	75		Si windows
Lazarus	0	4.7	5.1	88	0.8	21-124	0.49	Si windows
Sarek	1	10 (11)	6.7	1200	0.85	150-320	0.46	double
Lazarus	0	4.7 (6)	5.1	200	0.9	42/76		double, Si windows

Table 2. ^{3}He cells used for MEOP. All cells are valved, hence there is no column for pressure. Achievable polarization depended on operating conditions and can be found in the references provided. All cells were fabricated from blown GE180, with transitions to Pyrex stopcock valves, except where noted.

Cell name	d	l	V	T_1	References, notes
UNCOATED					
Mercury	4.4	10	155	37	[14, 59, 57], 1720 body, D1720 flat windows
Saturn	9.5	150	1060	50	[57], 1720 body, D1720 flat windows
Neptune	7.2	10.0	410	72	[57], 1720 body, D1720 flat windows
Venus	5.9	10	270	27	[60], 1720 body, GE180 ends
Earth	6.2	9	270	15	[61], 7056, O-ring valves
Earth-II	5.9	8.5	230	33	1720, O-ring valves
Mars	5.9	8.0	220	25	1720 body, GE180 windows
Rb-COATED					
Orion	7.5	10	440	240	[10, 30, 58]
Betelgeuse	7.5	10	440	230	[10]
Spica	7.0	10	390	74-258	[10, 4], see text
Antares	7.5	10	440	110	[4]
Arcturus	7.5	10	440	170	see text
Europa	5	5	100	30	[4]

Table 3. ^{3}He cells used for neutron scattering applications (SANS - small angle neutron scattering; REFL - reflectometry; TAS - triple-axis spectrometry; MACS - Multi-Axis Crystal Spectrometer). The MACS cells are discussed in Sec. 5.

Cell name	D	d	l	V	p	T_1	P_{He}	References, notes
SANS/REFL								
Snoopy	0	5.6	1.5	34	3.4	180	0.75	[59, 15]
Joe Cool	0	6.9	1.2	45	3.1	120	0.63	[10, 16, 15]
Bullwinkle	0	6.9	7.1	268	1.3	550	0.72	[16, 67, 68, 69, 15, 11, 66]
Chekhov	0	8.1	6.0	310	1.3	340	0.79	[4, 30, 11, 55]
Chardonnay	1.6	8.4	7.3	400	1.07	390	0.77	[15, 11, 55]
Orvieto	2.2	8.7	7.2	430	1.1	80	0.70	[15, 11, 40, 41, 42]
Olaf	4.4	10.8	6.0	545	0.96	348/472	0.72	[65]
Burgundy	2.2	12.0	8.0	895	1.09	450	0.80	[55, 63, 64]
Frascati	4.2	12.0	7.0	795	1.19	160	0.76	
Maverick	4	11.2	6.2	615	1.4	370		[63, 64]
Icemann	3.4	11.8	6.8	745	0.75	122	0.76	
Peabody	0	7.0	7.8	300	1.17	650	0.62	[69],sol-gel coating
(Yeoman) Rand	0	6.0	6.0	175	1.15	650		[69, 15]
(Nurse) Chapel	0	7.2	6.5	260	1.25	330	0.75	[30, 15]
TAS								
Zinfandel	1.0	11.6	8.9	940	1.86	330	0.78	[70, 11]
Syrah	6.2	10.2	9.7	790	1.43	450	0.76	[70, 11, 55]
Beaujolais	3.2	10.5	11.7	1010	1.37	336	0.69	
Sangiovese	3.4	11.0	10.0	950	1.76	354	0.79	
Chablis	7	11.3	9.4	945	1.65	187-270	0.75	
Mercurey	3.0	11.7	7.8	840	1.76	357	0.69	
MACS[7]								
Monastrell	≈5	8	6	300	2.2	160/190	0.70	polarizer
Gamay	≈2	7	5.5	170	2.1	90/120		polarizer
Grenache	5	7	5.5	170	2.35	140/220	0.63	polarizer
Barolo	11	8	11.5	580	1.3	170/190	0.7	30 ° analyzer
Barbaresco	11	8	11.5	580	1.3	250/320	0.7	30 ° analyzer
California	1	7	7	1100	1.35	17/32	0.4	120 ° analyzer (FQ plate)
Crescent	0	7	7	1100	1.32	30/50	0.4	120 ° analyzer (FQ plate)
Rainbow	0	7	7	1100	1.24	20/90	0.42	120 ° analyzer (FQ plate)
Jekyll	0	7	13	500	0.83	200	0.47	FQ analyzer (blown, test)
Peru	0	6.8	-	840	1.25	55/65		120 ° FQ analyzer (blown)
Jersey	0	7.8	-	1100	1.17	190	0.42	120 ° FQ analyzer (blown)

Journal of Physics: Conference Series **294** (2011) 012003 doi:10.1088/1742-6596/294/1/012003

Table 4. ^{3}He cells used or fabricated for neutron scattering applications at the Single Crystal Diffractometer (SCD) at Argonne National Laboratory or the magnetism reflectometer (MR) at the Spallation Neutron Source and for fundamental neutron physics applications (NPDGamma and interferometry experiments).

Cell name	D	d	l	V	p	T_1	P_{He}	References, notes
SCD/MR								
Felix	0	1.6	5.1	11	2.8	120	0.62	[15]
Oscar	0	1.7	5.1	12	3.1	20	0.51	[75, 15]
Marcy	0	3.5	5.3	51	2.2	360	0.67	[15, 76]
Colden	0	4.4	4.6	70	2.53	140/300	0.69	
Phelps	0	4.7	5.0	79	2.26	240	0.72	
Chianti	2.0	11.8	8.1	875	1.82	290	0.75	[70, 11]
Barbera	4.4	11.8	7.5	820	1.52	310	0.76	[70, 11, 77, 78]
Malbec	3.8	10.9	8.9	820	1.69	114/186	0.69	[21]
Bordeaux	3	11.4	8.2	840	1.09	80/310	0.67	
NPDGamma								
BooBoo	0	12.6	4.8	590	1.19	520		[10, 81]
Dino	0	10.6	5.1	450	0.9	530	0.75	[10, 16, 30, 15, 11, 81]
Pebbles	0	11.2	5.2	508	0.87	350	0.76	[10, 30, 15, 81, 40]
Kirk	0	10.5	7.2	620	0.80	740	0.77	[11, 81]
McCoy	0	9.7	6.1	450	1.16	470	0.66	
Yogi	0	10.6	4.9	432	0.9	180		[10]
Elroy	0	11	4.5	430	0.85	98		[10, 81]
Rocky	0	13.4	5.5	770	0.85	150		[81]
Astro	0	11.3	6.4	640	0.89	740	0.72	[10, 16, 30, 81]
Elmer	0	7 (5)	1.6	120	3	51	0.27	[79], double, meniscus NSF
Roadrunner	0	5 (4.5)	1.5	95	3	220		[42]
Interferometry								
Cashew	0	2.3	3.4	14	1.9	330	0.65	[15, 18, 19], flat windows
Peanut	0	2.3	3.4	14	1.7	8		[15, 18, 19], flat windows
Almond	0	2.3	3.4	14	1.8	35		[18, 19], flat windows
Pistachio	0	2.3	3.4	14	1.7	35		[18, 19], flat windows
Haystack	0	4	5.8	70	2.94	140	0.72	[18, 19], polarimetry
Skylight	0	4	5.8	70	3.1	156	0.69	[18, 19], polarimetry
Whiteface	1	4.5	5.9	94	3.50	28/143	0.74	[18, 19], polarimetry

References

[1] Walker T G and Happer W 1997 *Rev. Mod. Phys.* **69**, 629
[2] Bouchiat M A, Carver T R, and Varnum C M 1960 *Phys. Rev. Lett.* **5**, 373
[3] Babcock E *et al.* 2003 *Phys. Rev. Lett.* **91** 123003
[4] Jacob R E, Teter J, Saam B, Chen W C, and Gentile T R 2004 *Phys. Rev. A* **69** 021401(R)
[5] Ino T and Muto S 2007 *Physica B: Condensed Matter* **397** 182
[6] Colegrove F D, Schearer L D, and Walters G K 1963 *Phys. Rev.* **132**, 2561
[7] Fu C B *et al* accepted by *Physica B: Condensed Matter*; doi:10.1016/j.physb.2010.10.002.
[8] GE Lighting Component Sales, Bldg. 315D, 1975 Noble Rd., Cleveland, OH 44117. The largest standard GE180 tubing diameter is 16 mm; the 25 mm tubing was procured through a special glass run. Certain trade names and company products are mentioned in the text or identified in an illustration in order to adequately specify the experimental procedure and equipment used. In no case does such identification imply recommendation or endorsement by the National Institute of Standards and Technology, nor does it imply that the products are necessarily the best available for the purpose.
[9] Corning Glass, Corning, NY 14831
[10] Rich D R, Gentile T R, Smith T B, Thompson A K, and Jones G L 2002 *Appl. Phys. Lett.* **80**, 2210. In this reference cells were not identified by name, hence we provide this guide for cross reference: 1-Wilma, 2-Boris, 3-Barney, 4-Pebbles, 5-Dino, 6-Yogi, 7-Astro, 8-BooBoo, 9-Elroy, 10-Betty, 11-Natasha, 12-Spica, 13-Orion, 13-Betelgeuse
[11] Chen W C, Gentile T R, Walker T G, and Babcock E 2007 *Phys. Rev. A* **75** 013416
[12] Ocean Optics Inc., 380 Main Street, Dunedin, FL 34698.
[13] Ava-Spec-2048-2-SPV with VD-1800 grating (756 nm to 822 nm) for channel one and VE-2400 grating (393 nm to 482 nm) for channel two, Avantes Inc., 9769 W 119th Dr., Suite 4, Broomfield, CO 80021
[14] Jones G L 2000 *et al* Nucl. Instrum. Meth. **A440** 772
[15] Babcock E, Chann B, Walker T G, Chen W C, and Gentile T R 2006 *Phys. Rev. Lett.* **96** 083003
[16] Chann B *et al* 2003 *J. Appl. Phys.* **94** 6908
[17] Chen W C *et al* 2009 *Physica B: Condensed Matter* **404** 2663
[18] Huber M G *et al* 2009 *Nucl. Instrum. Meth. A* **611** 235
[19] Huber M G *et al* 2009 *Phys. Rev. Lett.* **102** 200401; Huber MG *et al* 2009 *Phys. Rev. Lett.* **103** 179903
[20] Spectraphysics diode bars, now sold by Oclaro Inc., 2584 Junction Ave. San Jose, CA 95134 USA
[21] Boag S *et al* 2009 *Physica B: Condensed Matter* **404** 2659
[22] Spectraphysics Comet (now sold by Oclaro Inc., 2584 Junction Ave. San Jose, CA 95134 USA) with Model 763H-FC mount and 3150 temperature controller from Newport Inc., 1791 Deere Avenue, Irvine, CA 92606 USA
[23] Optitemp Inc., P.O. Box 5246, Traverse City, Michigan 49696 USA.
[24] 80/20 Inc., 1701 South East, Columbia City, IN 46725 USA.
[25] Optigrate, Inc., 3267 Progress Drive, Orlando, FL 32826.
[26] We thank M.J. Barlow for information and assistance with the chirped grating approach.
[27] Abragam A 1961 *The Principles of Nuclear Magnetism* (Oxford University Press, Oxford, England).
[28] Bloch F 1946 *Phys. Rev.* **70** 460
[29] Newbury N R, Barton A S, Cates G D, Happer W, and Middleton H 1993 *Phys. Rev. A* **48** 4411
[30] Gentile T R *et al* 2005 *J. Res. Natl. Inst. Stand. Technol.* **110** 299
[31] Schearer L D and Walters G K 1965 *Phys. Rev.* **139** A1398-A1402
[32] Jacob R 2003 *Studies of hyperpolarized relaxation* ^{3}*He at glass surfaces* Univ. of Utah
[33] Deninger A, Heil W, Otten E W, Wolf M, Kremer R K, and Simon A 2006 *Eur. Phys. J. D* **38** 439
[34] Borel P I, Sogaard L V, Svendsen W E, and Andersen N 2003 *Phys. Rev. A* **67** 062705
[35] Masuda Y, Ino T, Skoy VR and Jones GL 2005 *Appl. Phys. Lett.* **87**, 053506
[36] O'Rourke Enterprises, 694 Main Street, P.O. Box 52, Lumberton, New Jersey, USA.
[37] Babcock E 2005 *Spin-exchange optical pumping with alkali-metal vapors* University of Wisconsin
[38] Lelievre-Berna E 2007 *Physica B: Condensed Matter* **397** 162
[39] Anger B 2008 *Polarization and relaxation mechanisms in hyperpolarized* ^{3}*He and* ^{129}*Xe* University of Utah
[40] Sharma M *et al* 2008 *Phys. Rev. Lett.* **101** 083002
[41] Babcock E *et al* 2009 *Physica B: Condensed Matter* **404** 2655
[42] Babcock E *et al* 2009 *Phys. Rev. A* **80** 033414
[43] Babcock E *IOP Journal of Physics Conference Series, this volume*
[44] Chupp T E and Loveman R A *Phys. Rev. C* **45** 915
[45] Milner R G, McKeown R D and Woodward C E 1989 *Nucl. Insrum. Meth. A* **274** 56

[46] Andersen K H *et al* 2009 *Physica B: Condensed Matter* **404** 2652
[47] Fitzsimmons W A, Tankersley L L, and Walters G K 1968 *Phys. Rev.* **179** 156
[48] Heil W *et al* 1995 *Phys. Lett. A* **201** 337
[49] Heil W *et al* 1999 *Physica B: Condensed Matter* **267-268** 328
[50] Jacob R E, Morgan S W, and Saam B 2002 *J. Appl. Phys.* **92** 1588
[51] Blue Flame Technology, Fairview, Texas, USA.
[52] Technical Glass Products, Painesville, Ohio 44077 USA.
[53] Romalis M V and Cates G D 1998 *Phys. Rev. A* **58** 3004
[54] Chann B, Babcock E, Anderson L W, and Walker T G 2002 *Phys. Rev. A* **66** 032703
[55] McIver J W, Erwin R, Chen W C and Gentile T R 2009 *Rev. Sci. Instrum.* **80** 063905
[56] Ye Q *et al* 2008 *Phys. Rev. A* **77** 053408
[57] Gentile T R *et al* 2001 *J. Res. Natl. Inst. Stand. Technol.* **106** 709
[58] Hussey D S 2005 *Rev. Sci. Instrum.* **76** 053503
[59] Gentile T R *et al* 2000 *J. Appl. Crystallog.* **33** 771
[60] Hussey D 2002 *Appl. Phys. A* **74** [Suppl.] S234
[61] Gentile T R *et al* 2000 *Magn. Reson. Med.* **43** 290
[62] Krycka K L *et al* 2009 *Physica B: Condensed Matter***404** 2561
[63] Krycka K L *et al* 2010 *J. Appl. Phys.* **107** 09B525
[64] Krycka K L *et al* 2010 *Phys. Rev. Lett.* **104** 207203
[65] Ueland B G *et al* 2010 *Phys. Rev. Lett.* **104** 147204
[66] Babcock E, Ioffe A, Radulescu A, Pipich V, arXiv:1006.4418.
[67] Chen W C *et al* 2003 *Physica B: Condensed Matter* **335** 196
[68] Chen W C 2004 *Rev. Sci. Instrum.* **75** 3256
[69] Gentile T R *et al* 2005 *Physica B: Condensed Matter* **356** 96
[70] Chen W C *et al* 2007 *Physica B: Condensed Matter* **397** 168
[71] Tian W *et al* 2008 *Phys. Rev. B* **78** 184429
[72] Chen Y *et al* 2008 *Phys. Rev. B* **78** 064515
[73] Lee S *et al* 2008 *Phys. Rev. B* **78** 100101(R)
[74] Cabrera I *et al* 2009 *Phys. Rev. Lett.* **103** 087201
[75] Jones G L *et al* 2005 *Physica B: Condensed Matter* **356** 91
[76] Jones G L *et al* 2006 *Physica B: Condensed Matter* **385-386** 1131
[77] Lauter V *et al* 2009 *Physica B: Condensed Matter* **404** 2543
[78] Lee W T *et al* 2010 *J. Physics: Conf. Series* **251** 012086
[79] Mitchell G *et al* S 2004 *Nucl. Instrum. Meth. A* **521** 468
[80] Gericke M T *et al* 2006 *Phys. Rev. C* **74** 065503
[81] Chupp T E *et al* 2007 *Nucl. Instrum. Meth. A* **574** 500
[82] Seo P-N *et al* 2008 *Phys. Rev. ST Accel. Beam* **11** 084701
[83] Gericke M *et al* 2009 *Nucl. Instrum. Meth. A* **611** 239
[84] Pokotilovski Yu N 2010 *Phys. Lett. B* **686** 114
[85] Thomas Gentile *et al*, http://meetings.aps.org/link/BAPS.2010.APR.D10.5
[86] Gentile T R and Fuller J August 2007 *Fusion (Journal of the American Scientific Glassblowers Society)* 21
[87] Wallidge G W *et al* 2005 *Physica B: Condensed Matter* **356** 118
[88] Parnell S R *et al* 2009 *Nucl. Instrum. Meth. A* **598** 774

Journal of Physics: Conference Series **294** (2011) 012004 doi:10.1088/1742-6596/294/1/012004

Structure of glasses for ^{3}He neutron spin filter cells

Y Sakaguchi[1, 9 *], H Kira[1], T Oku[1], T Shinohara[1], J Suzuki[1,9], K Sakai[1], M Nakamura[1], K Suzuya[1], K Aizawa[1], M Arai[1], M Takeda[1], Y Endoh[1], L J Chang[1, 2, 3], Y Arimoto[4], T Ino[4], H M Shimizu[4], T Kamiyama[4], K Ohoyama[5], H Hiraka[5], K Tsutsumi[6], K Yamada[7], K Ohara[8] and K Kakurai[1]

[1]Japan Atomic Energy Agency, Tokai, Naka-gun, Ibaraki 319-1195, Japan
[2]Department of Physics, National Cheng Kung University, Tainan 70101, Taiwan
[3]Nuclear Science and Technology Development Center, National Tsing Hua University, Hsinchu 30013 Taiwan
[4]Institute of Materials Structure Science, High Energy Accelerator Research Organization (KEK), Tsukuba 305-0801, Japan
[5]Institute for Materials Research, Tohoku University, Aoba-ku, Sendai 980-8577, Japan
[6] Department of Physics, Tohoku University, Aoba-ku, Sendai 980-8578, Japan
[7]WPI Advanced Institute for Materials Research, Tohoku University, Aoba-ku, Sendai 980-8577, Japan
[8]Japan Synchrotron Radiation Research Institute, SPring-8, Sayo 679-5198, Japan

E-mail: y_sakaguchi@cross.or.jp

Abstract. In spin-exchange optical pumping, there is unknown excess relaxation, called the *X*-factor, that limits attainable ^{3}He polarization. It is known that the *X*-factor depends on the glass containers. Hence, it is expected that there are cell-to-cell variations in microscopic structure among the cells and that these variations affect the attainable ^{3}He polarization. However, microscopic structure of the glasses for ^{3}He neutron spin filters has not been clarified yet. In this paper, we have performed precise X-ray diffraction measurements for the glasses at SPring-8 using synchrotron radiation. The structural differences are observed between several types of glasses, and also, between the GE180 glass pieces with different thermal treatment. Based on the results, the structural influence on the performance of ^{3}He neutron spin filters is discussed.

1. Introduction

Polarized ^{3}He neutron spin filters have attracted much attention as polarizing neutron devices for polarized neutron scattering experiments because of their wide coverage in neutron energy range and large acceptance in solid-angles. Due to the recent progress in the high ^{3}He polarization technique, they are now used as one of the options at neutron instruments in several neutron scattering facilities [1]. Spin-exchange optical pumping (SEOP) is one of the methods to polarize ^{3}He gas and its polarization can now reach about 80 % [2-4]. From an application point of view, it is desired to obtain much higher ^{3}He polarization. However, the attainable polarization seems to be saturated at about 80 %. In recent studies on the SEOP, it is reported that there is an unknown excess relaxation, called the X factor, that scales linear with alkali-metal density limiting the attainable ^{3}He polarization to $P_{He}<1/(1+X)$ [2, 5]. According to a number of measurements, the X factor

[9] Present address: Research Center for Neutron Science and Technology, Comprehensive Research Organization for Science and Society, Tokai, Ibaraki 319-1106, Japan

has surface to volume dependence that varies greatly from cell to cell [2]. Considering this fact, the microstructure of the glass cell could be an important parameter to help understand the nature of the X factor, although there could be other factors influencing the X factor such as purity of rubidium and cleanliness of the cell. In addition, it is empirically pointed that reblown glass has typically yielded the longest relaxation times [5, 6]. This can also support that the micro-structural difference affects the ^{3}He polarization. Probably, subtle differences in the glass fabrication process leads to a structural difference in the glass cells. However, as far as we know, there has been no prior study to show such differences from a microscopic point of view. Recently, we have performed X-ray diffraction measurements of glasses used for ^{3}He neutron spin filters, including GE180 [7]. GE180 is an aluminosilicate glass that is regarded as the most suitable material for SEOP cells. In this work we found that there could be a structural differences between GE180 pieces with different thermal treatments [8], but we could not make definite conclusions because of the accuracy of the measurements. In this paper, we have measured X-ray diffraction for several glasses for ^{3}He neutron spin filters using synchrotron radiation at SPring-8. Based on the results, the structural difference in the glasses is discussed.

2. Experimental

The high-energy X-ray diffraction experiments were carried out at the bending magnet beamline, BL04B2 [9] of SPring-8 with a two-axis diffractometer for disordered materials [10]. The incident photon energy of 61.7 keV, which was obtained from a bent Si (220) crystal, was used for the experiment. The measurements were performed in transmission geometry.

The samples that we measured are as follows; (1) vitreous silica, (2) BOROFLOAT [11], which is borosilicate glass and corresponds to Pyrex produced by Corning (3) GE180 and (4) PS-94 [12]. PS-94 is a lead-free glass and does not contain boron. Therefore, we thought that PS-94 could be a possible substitution for GE180. For the GE180 glass samples, we prepared four pieces with different thermal treatments; (1) a glass piece obtained from a glass tube after purchasing with no treatment (tube) (2) a blown GE180 glass piece, which was prepared for ^{3}He neutron spin filters and annealed at 780 °C for 30 min (blown) (3) a GE180 plate, which was molten and solidified in a mould and annealed at 750 °C for 20 min (plate, annealed). (4) a GE180 plate, which was molten and solidified in a mould, but without annealing (plate, not annealed).

3. Results and discussion

3.1. Structure factor

Fig. 1 shows the structure factor, $S(Q)$, for several types of glasses. We could obtain the structure factor in the Q-range from 0.16 Å^{-1} to 25.3 Å^{-1} with good accuracy; in previous measurements using an in-house X-ray diffractometer, the covered Q-range was from 0.4 Å^{-1} to 7.0 Å^{-1} and we could only observe the first peak and the other peaks could not be recognized. Comparing the data of GE180 with other three glasses, the features at 2.9 Å^{-1} and 6.5 Å^{-1} are clearly different. This could be a sign of the appearance of a new local structural order produced by the addition of other elements such as aluminium to the original SiO_2 matrix. As seen in Fig. 1(b), a difference in the position of the first sharp diffraction peak (FSDP) is also observed. This is consistent with our previous result [8]. The FSDP indicates the presence of a medium-range order with about 4.2 Å (for Q=1.5 Å^{-1}) and it is interpreted as an indicative peak of either the cages formed by the topological connection of tetrahedral $Si(O_{1/2})_4$ units [13] or voids [14]. Assuming the void model, the peak position indicates the size of the microscopic holes. Such holes are responsible for helium permeation that can cause ^{3}He depolarization [15]. Since the FSDP position of GE180 is higher (Q=1.87 Å^{-1}) than that of vitreous silica (Q=1.53 Å^{-1}), the helium permeation constant of GE180 is expected to be less than that of vitreous silica. Also, the FSDP position of PS-94 (Q=1.96 Å^{-1}) is higher than that of vitreous silica. As far as the size of microscopic holes is concerned, PS-94 seems to be suitable for ^{3}He neutron spin filter cells.

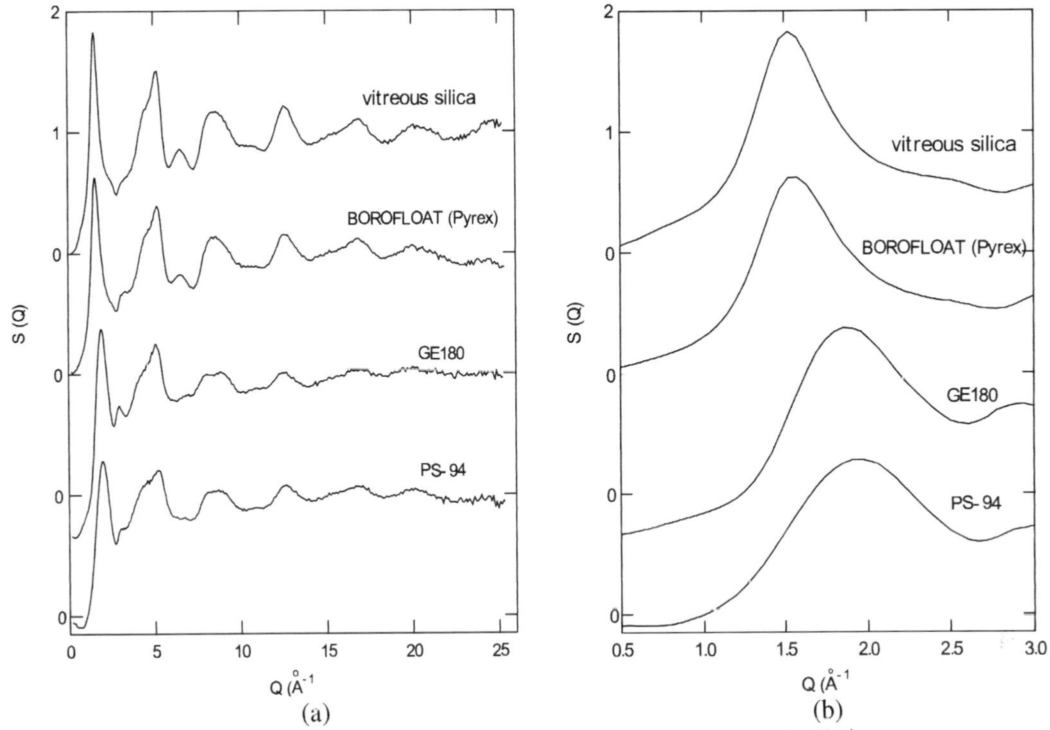

Figure 1. Structure factor, $S(Q)$, for several types of glasses. (a) $S(Q)$ measured over the whole Q-range (b) $S(Q)$ in the range from Q=0.5-3.0 Å^{-1}.

In Fig. 2, the correlation between the position of FSDP and helium permeation velocity constant, K, at 100 °C [16] is shown. To plot the point of borosilicate, the FSDP position of BOROFLOAT and the helium permeation velocity constant of Pyrex are used. To plot the point of aluminosilicate, the FSDP position of GE180 and the helium permeation velocity constant of Corning 1720 are used [17]. This good correlation confirms our expectation that the position of the FSDP indicates the size of the microscopic holes in the glass network.

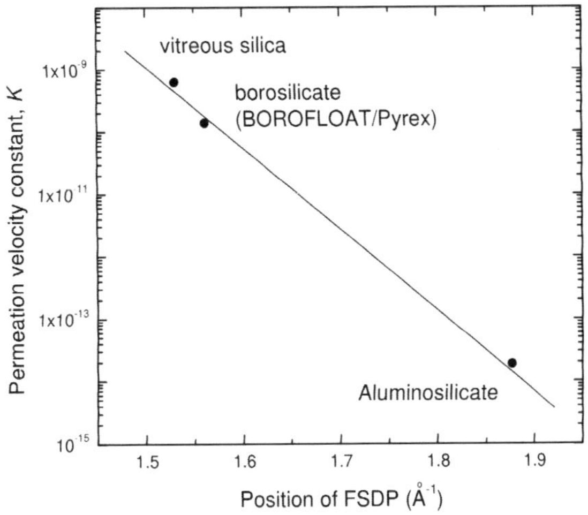

Figure 2. Correlation between the position of the FSDP and helium permeation velocity constant at 100 °C [16].

Fig. 3 shows the structure factor for several GE180 pieces with different thermal treatment. The curves almost overlap each other. However, the feature of blown GE180 is a little bit different from the others; a decrease in intensity from 5 to 17 Å^{-1} and a difference in symmetry of the peak at 20 Å^{-1}. Such change would be related to the changes in the short-range order, rather than the medium range. In the previous measurement, we have pointed out that the position of the FSDP can be different among those GE180 pieces. However, we could not find a clear difference in the position in the present work as shown in the inset.

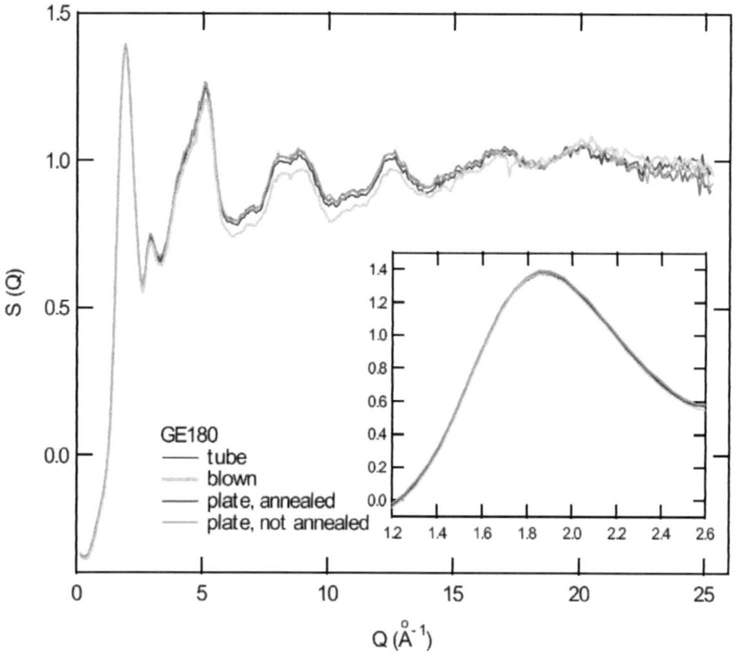

Figure 3. Structure factor, $S(Q)$, for several GE180 pieces with different thermal treatment.

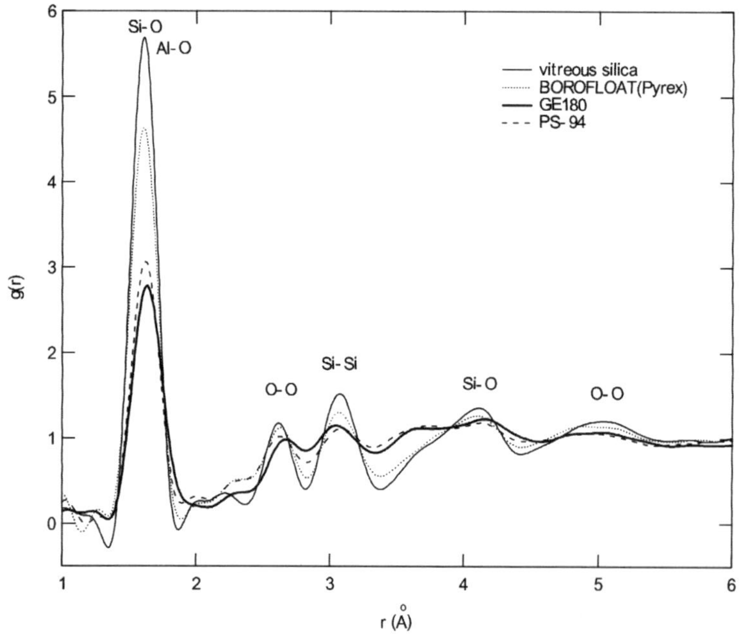

Figure 4. Pair distribution functions, $g(r)$, of several types of glasses.

Journal of Physics: Conference Series **294** (2011) 012004 doi:10.1088/1742-6596/294/1/012004

3.2. Pair distribution functions

Fig. 4 shows the pair distribution functions, $g(r)$, for several types of glasses. The peaks from the first to the fifth for vitreous silica are related to the first Si-O bond, the first O-O bond, the first Si-Si bond, the second Si-O bond and the second O-O bond, respectively. The peak positions are consistent with previously measured results for vitreous silica [18]. For BOROFLOAT and PS-94, the positions of the first, the second and the third peaks are basically the same as that for vitreous silica. This means that the network structure composed of Si-O bonds is preserved and the constitutional unit of $Si(O_4)_{1/2}$ tetrahedron is also preserved in the glasses. For GE180, the first peak is a little bit asymmetrical and it is considered that a new peak appeared at the longer-r side, which is attributed to Al-O bonds [19].From double-Gaussian fitting, the position was estimated to be 1.74 Å. Although the positions of the second and the third peaks are not exactly the same as those for vitreous silica, the structure of $Si(O_4)_{1/2}$ tetrahedral unit is supposed to be preserved in GE180. The difference in this region would arise from appearance of new local order due to the addition of other element like aluminium.

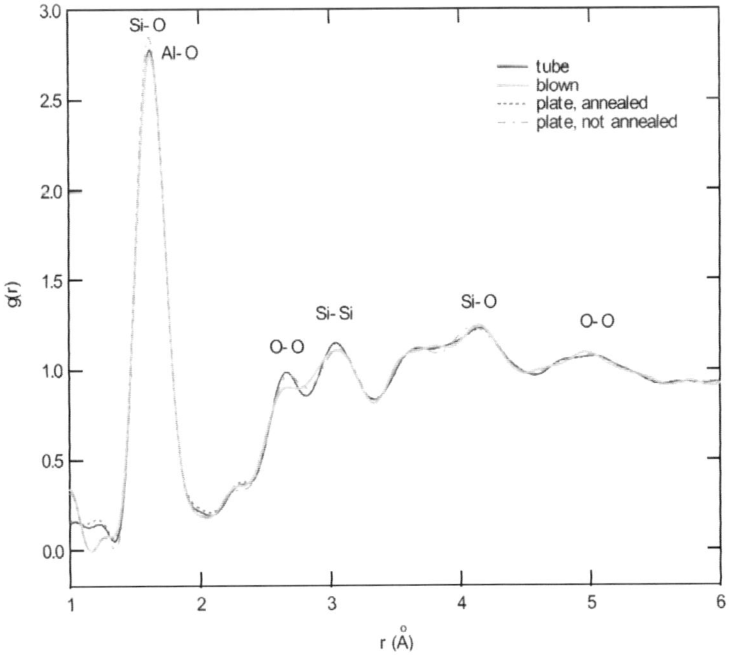

Figure 5. Pair distribution functions, $g(r)$, of several GE180 glass pieces.

Fig. 5 shows the pair distribution functions, $g(r)$, for several GE180 pieces with different thermal treatment. The position and the height of the first peak are almost the same among the four GE180 glass pieces. However, there is a clear difference in the second and the third peaks between the blown GE180 glass piece and the others; the second and the third peaks for the blown GE180 piece are more broadened. This indicates that the bond angles of Si-O-Si and O-Si-O in the blown glass are more widely distributed than the others, in other words, the tetrahedral units are more distorted from perfect tetrahedron in the blown GE180 glass. The increase of such structural randomness by the blowing process is plausible because external force was applied in molten condition (mobile for atoms) and the more disordered status was quenched. As the FSDP result in
Fig. 3 indicates, the average size of the microscopic holes is considered to be almost the same. Interestingly, these situations are very similar to those in high temperature vitreous SiO_2 [20]. In the real space, the first peak does not change and the second peak decreases. In Q-space, the FSDP does not change. From the similarities, the structure of blown GE180 glass can be

regarded as a quenched state with more disordered structure in the molten state. Considering the cages, which are frameworks of microscopic holes, two different cages are orthogonally linked and centred at the Si atom in one tetrahedral $Si(O_{1/2})_4$ unit. When the bond angles become flexible the angles between the two cages also become flexible. This will lead to more complicated three-dimensional arrangement of the cages in the glass. Hence, it is expected that helium atoms do not pass through the microscopic holes as easily, and that the local helium permeation near the glass surface is suppressed.

3.3. Structural consideration on ^{3}He neutron spin filter cells

In order to find what the unknown excess relaxation process is, it would be helpful to perform a study on the worst scenario because the changes due to the process would be maximized there. Among the glasses used for ^{3}He spin filters, vitreous silica is known as a glass with high X factor [2], in which the excess relaxation is greatly affected. For such vitreous silica cells, Ino et al. have obtained long relaxation time close to the dipole-dipole limit [21] using the blown cells [22]. We infer that similar situation occurs in vitreous silica to that in GE180; helium permeation is decreased by blowing the glass and the situation must have minimized the influence of magnetic impurities in the glass. During optical pumping, there should be a different process. Because of the size of the microscopic holes, helium atoms can permeate the glass while rubidium atoms cannot. In this situation, unpolarized helium atoms in the glass do not have a chance to polarize by colliding with polarized rubidium atoms. Presence of such unpolarized helium atoms would limit the attainable ^{3}He polarization. Actually, the vitreous silica cells, which have large microscopic holes, have lower attainable ^{3}He polarization [22-24].

In the present study, it was found that blown GE180 glass has more disordered structure. From the result, we inferred that the local helium permeation near the glass surface is suppressed. This could work to reduce the content of helium atoms which do not have a chance to collide with polarized rubidium atoms, and hence to get higher attainable ^{3}He polarization. Such possibilities have to be examined by further analysis on the structure of the glasses used for ^{3}He spin filters.

4. Summary

We have performed precise X-ray diffraction measurement of several glasses for ^{3}He neutron spin filter cells using synchrotron radiation in SPring-8. From the measurement, it was found that the FSDP can be a good measure to estimate the size of microscopic holes through which helium atoms can pass. More disordered structure has been observed in blown GE180 glass. This structural change would give rise to the decrease in the local helium permeation near the glass surface, and thus, lead to higher ^{3}He polarization and longer relaxation times. The empirical finding that blown GE180 glass had typically yielded the longest relaxation times is considered to be related to such microscopic structural changes.

Acknowledgments

This research project has been supported by Quantum Beam fundamental Development program, MEXT. The authors would like to thank Dr. S. Kohara for valuable discussion on the X-ray diffraction measurements.

References

[1] For example, Chen W C, Erwin R, Mclver III J W, Watson S, Fu C B, Gentile T R, Borchers J A, Lynn J W and Jones G L 2009 *Physica B: Condensed Matter* **404** 2663; Andersen K H, Jullien D, Petoukhov A K, Mouveau P, Bordennave F, Thomas F and Babcock E 2009 *Physica B: Condensed Matter* **404** 2652.

[2] Babcock E, Chann B, Walker T G, Chen W C, Gentile T R 2006 *Phys. Rev. Lett.* **96** 083003

[3] Parnell S R, Babcock E, Nünighoff K, Skoda M W A, Boag S, Masalovich S, Chen W C, Georgii R, Wild J M and Frost C D 2009 *Nuclear Instruments and Methods A* **598** 774
[4] Ioffe A and Babcock E 2011 *IOP journal of physics conference series, this volume*
[5] Chann B, Babcock E, Anderson L W, Walker T G, Chen W C, Smith T B, Thompson A K and Gentile T R 2003 *J. Appl. Phys.* **94** 6908
[6] Rich D R, Gentile T R, Smith T B and Thompson A K 2002 *Appl. Phys. Lett.* **80** 2210
[7] GE Lighting Component Sales, Bldg.315D, 1975 Noble Rd., Cleveland, OH 44117
[8] Sakaguchi Y, Kira H, Oku T, Shinohara T, Suzuki J, Sakai K, Nakamura M, Suzuya K, Arai M, Takeda M, Wakimoto S, Yamazaki D, Koizumi S, Endoh Y, Kakurai K, Arimoto Y, Ino T, Shimizu H M, Kamiyama T, Ohoyama K, Hiraka H, Tsutsumi K, Yamada K and Chang L J *Nuclear Instruments and Methods A* **634** Supplement 1 S122
[9] Isshiki M, Ohishi Y, Goto S, Takeshita K, Ishikawa T 2001 *Nuclear Instruments and Methods A* **467-468** 663
[10] Kohara S, Suzuya K, Kashihara Y, Matsumoto N, Umesaki N and Sakai I 2001 *Nuclear Instruments and Methods A* **467-468** 1030
[11] This is borosilicate glass produced by SCHOTT and is known as TEMPAX Float in Japan
[12] Nippon Electric Glass Co., Ltd
[13] Wright A C 1994 *J. Non-Cryst. Solids* **179** 84
[14] Elliott S R 1991 *Phys. Rev. Lett.* **67** 711
[15] Fitzsimmons W A, Tankersley L L and Walters G K 1969 *Phys. Rev.* **179** 156
[16] Norton F J 1953 *J. Am. Ceram. Soc.* **36** 90
[17] Although Corning 1720 contains boron oxide with 5 wt. % [16] while GE180 contains no boron, boron oxide acts as a network former as well as silicon oxide while aluminium oxide acts as a modifier. According to the result in [16], the content of the network former correlates with the helium permeation velocity constant. Since the helium permeation velocity constant for GE180 (SiO_2 ~60 wt. %) is expected to be close to the value for Corning 1720 (SiO_2 + B_2O_3= 67 wt.%) in the correlation, we used the value to plot in Fig.2.
[18] Kohara S and Suzuya K 2002 *Phys. Chem. Glasses* **43C** 51
[19] Petkov V, Billinge S J L, Shastri S D and Himmel B 2000 *Phys. Rev. Lett.* **85** 3436
[20] Kohara S and Suzuya K 2003 *Nuclear Instruments and Methods A* **199** 23
[21] Newbury N R, Barton A S, Cates G D, Happer W and Middleton H 1993 *Phys. Rev.* A **48** 4411
[22] Ino T, Muto S, Masuda Y, Kim G and Skoy V R 2005 *Physica* B*: Condensed Matter* **356** 109
[23] Fu C B, Gentile T R, Jones G L, Chen W C, Erwin R, Watson S, Broholm C, Rodriguez-Rivera J A and Scherschligt J 2011 *Physica B-Condensed matter* (Corrected Proof, Available online 6 October 2010) ISSN 0921-4526, DOI: 10.1016/j.physb.2010.10.002
[24] Chen W C, Gentile T R, Fu C B, Watson S, Jones G L, McIver J W and Rich D R 2011 *IOP journal of physics conference series, this volume*

Journal of Physics: Conference Series **294** (2011) 012005 doi:10.1088/1742-6596/294/1/012005

SEOP program for neutron spin filters at the Jülich Centre for Neutron Science

E Babcock*, **A Ioffe**

Jülich Centre for Neutron Science at FRM II, Institut für Festkörperforshung, Forschungszentrum Jülich GmbH, 85747 Garching, Germany

E-mail: e.babcock@fz-juelich.de

Abstract. The Jülich Centre for Neutron Science is developing spin-exchange optical pumping systems capable of polarizing the ^{3}He gas for use as neutron spin filters with developments focused on individually optimized solutions for each application. When possible steady polarization achieved with in-situ optical pumping will provide high time averaged neutron performance and time-stability, however some applications are foreseen to use a "local filling station" approach. In-situ polarizers are being developed for polarization analysis on reflectometry and small angle neutron scattering instruments. This overview of the current status of our SEOP program describes the motivation of in-situ polarization, the design and first results from the polarizer for analysis on reflectometry, and discusses the designs for a SANS analyzer and plans for wide angle analysis.

1. Introduction: Polarized ^{3}He used for neutron spin analysis

Polarized ^{3}He has been shown to be advantageous when used as an analyzer for polarized neutron scattering, an area of physics research that is growing quickly. These sorts of experiments are normally conducted in large scale facilities for neutron scattering research where even at the highest flux neutron sources these experiments can require long data collection times. This is partly because of the losses in neutron counting rates caused by the addition of neutron polarizers and analyzers, but also because one now normally must measure two to four combinations of the incident beam polarization and analyzer polarization, essentially performing the same experiment several times. Further factors such as small sample size in the case of magnetism studies or low angular coverage of the scattered beam further lengthen these types of experiments.

Much work has been done to optimize the use of polarized ^{3}He for use as neutron spin filters (^{3}He NSF). Routinely experiments using ^{3}He NSF are conducted at many neutron sources worldwide. These experiments include polarized neutron diffraction, polarized neutron reflectometry, neutron interferometry, polarized small angle neutron scattering and fundamental particle physics [1–6].

Two methods are typically used to polarize ^{3}He gas, spin-exchange optical pumping (SEOP) and metabstable-exchange optical pumping (MEOP) [7, 8]. In recent years both methods have been achieving ever higher understanding of the physical processes limiting the magnitude and rates of ^{3}He polarization. This along with technological achievements have produced ^{3}He polarizations of up to 80% in ^{3}He NSF cells that have been verified with neutron measurements

[1, 9–11]. These levels of performance make ^{3}He NSFs highly applicable in many areas of polarized neutron instrumentation.

This overview of the current status of the Jülich Centre for Neutron Science (JCNS) SEOP program will describe the motivation of in-situ polarization, the design and first results from our polarizer designed for reflectometry/grazing incidence small angle neutron scattering (GISANS), the designs for a soft matter SANS analyzer currently under construction, and instrumentation planning to use the "local filling station" approach.

2. Advantages of in-situ ^{3}He polarization

The polarization of ^{3}He gas, being an atomic physics process, is normally performed in the controlled conditions of an atomic-molecular-optical physics laboratory because conditions involving high power class IV lasers, highly uniform magnetic fields and isolation from electronic and magnetic interference are required. However ^{3}He polarization is a dynamic process involving spin-polarization and T_1 decay of nuclear spin, thus once the ^{3}He is polarized and in an appropriate container, and the polarization from optical pumping is no longer taking place, it begins to undergo T_1 relaxation to 0 polarization. This polarization decay places certain limits on experiments using off-line polarized gas.

In practical terms, T_1 lifetimes of 100 hours to 300 hours are the achievable goal in neutron experiments using ^{3}He NSFs. The neutron energy dependent absorption cross section for ^{3}He is nearly entirely spin dependent and the neutron polarization efficiency P_n is related to the ^{3}He polarization, P_{He}, and the product of the ^{3}He density-length and neutron cross section, Θ, by $P_n = tanh(P_{He}\Theta\lambda_n)$ where λ_n is the neutron wavelength. The increase in neutron transmission relative to the unpolarized transmission of the ^{3}He NSF is given by $T_n/T_0 = cosh(P_{He}\Theta\lambda_n)$ where T_0 is $e^{-\Theta\lambda_n}$. Since the neutron spin analyzing efficiency or quality can be shown to be on the order of $P_n^2T_n$ [12, 13] when the ^{3}He is allowed to undergo T_1 decay of polarization its quality as an NSF will also continually decay with time. Thus we can see the importance of maintaining as high a degree of time-averaged ^{3}He polarization as possible.

To take advantage of the full potential performance of a ^{3}He NSF one must actively counteract the ^{3}He T_1 decay trough periodic refreshing of the polarized gas or in-situ polarization. Commonly gas is polarized either with SEOP or MEOP in a laboratory away from the neutron instrument and then transported to the location of use. In this case, depending on the on instrument ^{3}He relaxation time, the cells are then exchanged every one to two days, with calibration measurements of the instrument necessary for each cell, or cell filling, to correctly normalize to the time dependence of the ^{3}He polarization [14, 15]. Additionally, refreshing of the polarized gas periodically in a pulsed mode has been proposed to maintain polarizations closer to those achieved in laboratory conditions in a pseudo-steady state. However, the only way to truly obtain the full potential for a ^{3}He NSF, is to develop a system capable of polarizing the ^{3}He gas in place, maintaining it at the steady state maximum polarization one obtains in the controlled laboratory conditions. In this case one further benefits by having time stability of the instrument.

We will assume that $P_n^2T_n$ is the relevant quality factor for a fixed neutron wavelength [12, 13] and that since both P_n and T_n are functions of ^{3}He polarization, the achieved instrument performance or quality factor Q will be proportional to the time integral of $P_n^2T_n$. When normalized to flux, Q can be thought of as the fraction of neutrons recorded on the the detector that can be used to obtain counting statistics. Assuming a neutron polarizer and polarization analysis (PA) system consisting of an ideal supper mirror (SM) incident beam polarizer (i.e. neutron transmission of 50% and neutron polarization of 100% for the unpolarized incident beam through the SM) and a ^{3}He analyzer operating at high ^{3}He polarization and optimal $P_n^2T_n$ ($P_n^2T_n \simeq 0.2 - 0.25$ typically) one would need to count for 4-5 times as long as without PA to obtain the same counting statistics. In addition to the lowering of count rates caused by the

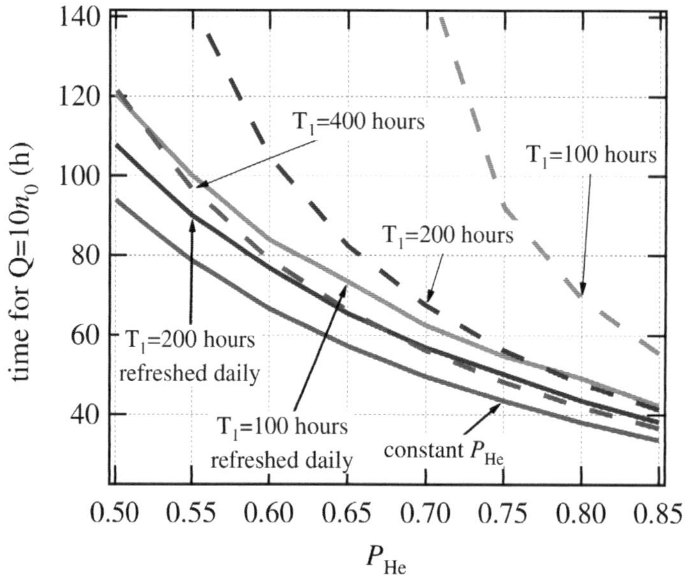

Figure 1. Amount of time required to obtain a quality of $10n_0$ for various conditions of steady and time decaying ^{3}He polarization. The solid red line is constant polarization, the solid blue/green line is time decaying polarization with a 200/100 hour lifetime that is refreshed to maximum polarization daily with a 1 hour pause in data for calibration and cell changing, the dotted grey/blue/green line is time decaying polarization with a 400/200/100 hour time constant. One can see that for any given attainable level of polarization, maintaining steady polarization will reduce measurement times for for a fixed statistical accuracy.

addition of PA, experiments using PA commonly measure the four possible combinations of the polarizer and analyzer spin states. Thus with PA one would measure a total of 16 to 20 times as long to obtain similar statistical errors for each of the four states as one would require for a single unpolarized measurement.

To show the effects of the the time dependence of Q we have created plots of the time required to measure $Q = 10n_0$ as a function of ^{3}He polarization for several different cases of time decaying and steady ^{3}He polarization. Here we define n_0 as the number of counts recorded in one hour from the scattered beam without PA installed on the instrument. Then the time to reach $Q = 10n_0$ would be the time required to record $2.5n_0$ counts for each of the four states giving a statistical accuracy of $1/(\sqrt{2.5n_0})$ for each state if one sample/set of conditions were measured. We simply note that n_0 as defined above is a function of the incident beam intensity, the neutron detector, and the neutron scattering cross section of the sample, thus the amount of time required to obtain a relevant level of statistical accuracy is highly variable and is dependent on all of these parameters. But for example, if one has a detector count rate of 111 Hz for the scattered beam without PA, it would take 2.5 hours to measure 1×10^6 counts, i.e. to obtain a statistical accuracy of 0.1%. Thus given our definition of n_0, $Q = 10n_0$ would correspond in time to the time required to obtain 1×10^6 counts in each of the four states if the detector count rate for the unpolarized measurement on the same sample with the same instrument were 111 Hz. Or for another example, to perform 10 measurements on a series of samples/sample conditions, when the count rate is 1.11 kHz without PA installed, and maintain statistical errors of 0.1% for all four states with PA.

Figure 1 thus shows how the quality factor Q evolves as a function of the maximum achievable ^{3}He polarization. One can see that to obtain Q=10n_0, one must count for about 38 hours with a constant P_{He}= 80% and about 50 hours with time decaying P_{He} starting at 80% with a 100 hour T_1 that has the polarization refreshed daily. 50 hours is the same time as required to obtain this level of statistical accuracy using a cell polarized constantly to P_{He} = 70%. A cell with a 100 hour T_1 initially polarized to P_{He} = 80% that does not have the polarization refreshed would require nearly double the counting time, about 70 hours, as the cell continuously polarized to 80% to reach Q=10n_0.

Also shown in figure 1 is a cell with a 400 hour T_1, a value similar to what we have obtained in tests on a SANS instrument discussed in Ref. [16]. Here the difference in time integrated quality becomes less over time scales short compared to T_1. As a rule of thumb, one can use a ^{3}He NSF cell for approximately one day per 100 hours of on-beam T_1. In this case, over long time scales, one would obtain a relative time averaged quality, $\bar{Q}/n_0$, about 3/4 of that obtained with constant polarization. This ratio can be seen by comparing the green and red line in Fig. 1 and is relatively independent of starting polarization or T_1 in limit where the measurement time is longer than the time one can use each cell filling, i.e. one day per 100 hours of T_1. Clearly one could chose to refresh the polarization more often to approach the levels obtained with constant polarization if the resources are available, as can be seen comparing the blue line to the red line in Fig. 1, however time dependent corrections and calibrations of each cell's polarization would still be required. From this one can see the advantage of higher time averaged polarization, and the optimal advantage of in-situ optical pumping. Because of this gain in performance, plus the additional advantages of time-stability and lower daily maintenance, first work at the JCNS will focus on the development of in-situ ^{3}He polarizers using the SEOP method.

3. SEOP ^{3}He Polarizer developed for a reflectometry polarization analyzer

The polarizer we developed for the new JCNS magnetism reflectometer, called MARIA, allows us to take advantage of the positive traits of an in-situ ^{3}He polarizer [17]. The MARIA reflectometer has a large area, 40 cm by 40 cm position sensitive detector that makes a ^{3}He spin filter cell an ideal choice for analysis, in accord with instrumentation developed at other centers for reflectometer applications [1, 14, 18, 19].

The polarizer device has been described previously in [11] and is shown in Fig. 2. The system is based on a "magic box" style magnetostatic cavity which provides a wide aperture opening for the scattered neutron beam and optical access perpendicular to the axis of neutron propagation through the system. Further, the μ-metal of this magnetic cavity provides good shielding from external magnetic fields and thus gives stable ^{3}He T_1 relaxation times independent of external factors or instrument settings.

Since we polarize the ^{3}He gas in-situ, we are able to achieve high stability and time averaged performance of the ^{3}He NSF. This device recently maintained a high level of ^{3}He polarization of 80.4%±1.5% in test measurements using the cell called J1 which has a diameter D=6 cm and is optimized for a cold neutron spectrum [11]. Such levels of polarization maintained over time provide high ^{3}He NSF performance for this application, and the stability over time simplifies data analysis and instrument calibration.

When used with a larger diameter cell, D=9 cm, this device was capable of maintaining the same saturation polarization as when this cell is polarized in laboratory conditions. For over 4 days during a polarized GISANS experiment we maintained a ^{3}He polarization of 73% while using the sample magnet and performing routine AFP spin reversal of the ^{3}He polarization to be discussed below. In this experiment we used the cell called Orvieto on loan from NIST [20], that provided approximately 5 degrees of angular coverage in the configuration of this particular experiment. During the week of operation the system allowed full flexibility of the MARIA sample environment and sample magnet which was operated at up to 1.2 T without causing any

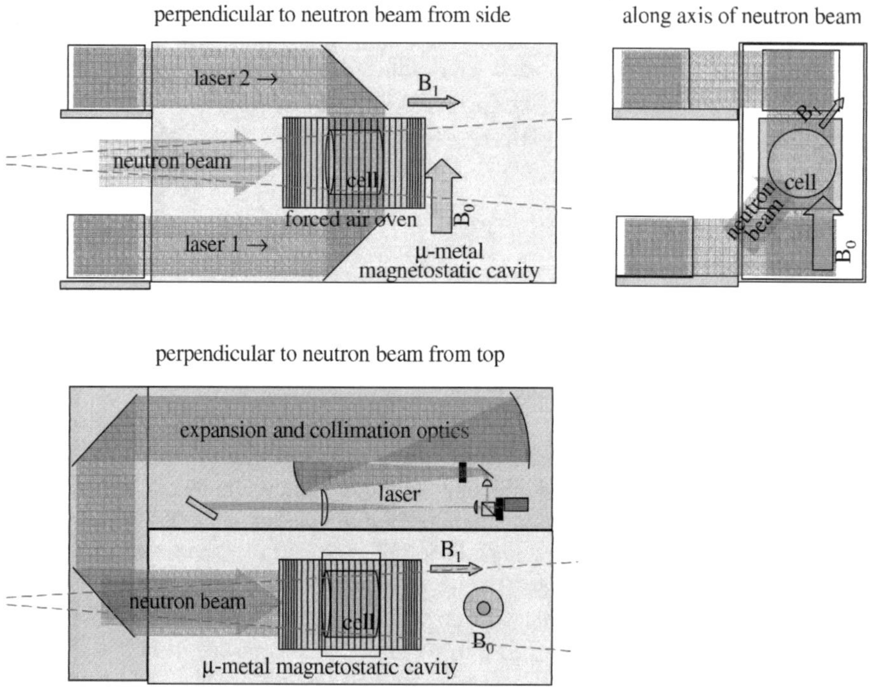

Figure 2. The ^{3}He polarizer developed for the MARIA magnetic reflectometer. This polarizer is based on a 76 cm long magic box style magnetostatic cavity with a 40 cm high by 30 cm wide cross section. This dimension allows full acceptance of the scattering angle allowed by the MARIA detector and laser access perpendicular to the scattered neutron beam for the optical pumping. The lasers are two frequency narrowed diode array bars that are collimated and expanded as shown in the diagram. They are then directed into the cell with mirrors.

noticeable effect on the ^{3}He polarization, or on the measured cell T_1 in the absence of optical pumping. The sample magnet on MARIA is a yoked electro-magnet, the return field of which couples weakly to the B_0 holding field of our magic box, changing the Larmor frequency of the ^{3}He atoms by about 50 Hz when its field is varied from 0 T to 1.2 T. During this experiment the ^{3}He cell was about 100 cm away from the center of the magnet bore, i.e. the front of the magic box was about 60 cm away from the center of the magnet and the ^{3}He cell was centered longitudinally in the 76 cm long magic box. A photo of the in-situ polarizer is shown in Fig. 3.

Another feature of this in-situ polarizer is to use the ^{3}He NSF as an ideal wide area neutron spin filter by performing adiabatic fast passage (AFP) spin reversal of the ^{3}He gas polarization. The AFP system is similar to that described in Ref. [21]. However for this application the AFP must also be synchronized to circular polarization reversal of the optical pumping lasers which we accomplish by using liquid crystal variable wave retarders whose control is written into the program that performs the AFP RF pulse generation discussed below. This allows us to perform continual optical pumping of the ^{3}He gas while the ^{3}He NSF spin state, or analyzing direction, is controlled by the MARIA instrument with a simple TTL interface.

Our AFP system is based on a digital 1.2 MHz computer data acquisition card and a high power (200 W), high voltage (± 75 V_{pp}), power amplifier driving an off-resonant solenoid coil

Figure 3. A picture of the ^{3}He polarizer developed for the MARIA magnetic reflectometer. When in operation the entire frame is covered with aluminum plates for laser radiation security. These outer aluminum plates have been removed to show the mounting of the laser optics with respect to the magic box.

wound directly onto the SEOP oven to provide the transverse RF field. A digital waveform is generated by a computer program that allows us to easily change the parameters. The program produces a frequency chirped wave from 0.25 to 1.5 times the Larmor frequency of the ^{3}He atoms at the chosen B_0 holding field which is normally about 10 gauss (corresponding to a Larmor frequency of 32 kHz). This chirped wave has a duration of 0.5 s and is amplitude modulated with a gaussian profile of width equal to 1/8 of the pulse duration to insure a fully adiabatic reversal of the ^{3}He. The house built power amplifier that allows us to drive a non-resonance coil is important because we can freely chose any B_0 field without modifying the coil tuning or circuit impedance matching as would be required with a resonant coil. The 0.5 s long sweep helps make the AFP robust against small changes in the B_0 holding field caused by slight field drifts due to power supply stability or coupling of the SEOP system's magnetic cavity to the sample magnet.

4. Design of an ultra compact polarization analyzer for SANS and plans for wide angle analysis on TOPAS and DNS

While the MARIA system focuses on coverage of a wide angle scattered beam in conjunction with a high sample field, there are other applications that will not require the use of high sample fields and thus can be made more compact, placing the cell very near to the sample in order to obtain high angular coverage with a relatively small cell and less laser power. Separation of incoherent background for soft-matter studies with small angle neutron scattering is one such application [22]. Initial testing of an off-line polarized cell showed the feasibility of this technique for the JCNS. Consequently we have designed and begun prototyping an ultra compact SEOP polarizer to be used for SANS. The system will use an 18 cm long μ-metal end-capped and shielded solenoid. The system will have a small entrance aperture and a large exit aperture in the end caps for the scattered neutron beam. The effects of these holes in the μ-metal end caps on the field homogeneity are then corrected with additional coils. This design will allow the sample to be placed within 3-4 cm of the ^{3}He cell while using standard sample environment.

Ultra-wide angle applications, i.e. involving angular coverage >90 degrees in the horizontal and or vertical planes, such as DNS [23] or the future TOPAS spectrometer [24], are foreseen to use a filling station approach. The possible system for DNS could be similar to the PASTIS approach [25] and use ^{3}He gas polarized off line which is transported to a single wide-angle cell in the DNS detector housing using the local filling technique [14]. The system would have the ^{3}He NSF close to the sample and given the cold energy neutrons, i.e $\lambda \simeq 4.5$ Å, it would not require more than 1-2 bar liters of polarized ^{3}He per cell filling. This system could serve as a prototyping stage for the much more demanding application on the TOPAS specterometer currently under construction. TOPAS will use thermal neutrons with wavelength $\lambda \simeq 1$ Å requiring on the order of 30 bar-cm of gas thickness in the NSF and an array of cells with a surface area of 0.7 m^2 to obtain full detector coverage. This extremely large volume of polarized ^{3}He will necessitate the development of new techniques and larger amounts of polarized gas per day than what is currently in use at any neutron research center. The polarizer will most likely be a large scale SEOP polarization station designed to have a throughput of 100 bar liters per day. This large scale station would then be connected to the array of cells via polarization preserving transport tubes and an automated vacuum pump/gas recycling system as in the local filling approach.

5. Summary

The ^{3}He spin filter program at the JCNS is making steady progress towards application of ^{3}He neutron spin filters on a variety of neutron instruments. Initial devices have been constructed and tested with positive results. Further developments will focus on refining the prototypes and making new novel designs customized to the particular application.

6. Acknowledgements

We kindly thank the continual collaboration and useful discussions with colleagues T. R. Gentile, W. C. Chenn and C. B. Fu at NIST; S. Boag and C. D. Frost at ISIS; S. R. Parnell at U. of Sheffield; and T. G. Walker at U. Wisconsin Madison. We additionally acknowlegde the continual involvement of S. Mattauch (Jülich Centre for Neutron Science) in our polarized ^{3}He efforts especially with respect to reflectometry application.

References

[1] Chen W C, Erwin R, McIver-III J W, Watson S, Fu C B, Gentile T R, Borchers J A, Lynn J W and Jones G L 2009 *Physica B-Condensed matter* **404** 2663–6
[2] Ino T, Nakamura M, Oku T, Shinohara T, Suzuki J, Ohoyama K and Hiraka H 2009 *Physica B-Condensed Matter* **404** 2667–9
[3] Boag S, Parnell S R, Frost C D, Andersen K H and Babcock E 2007 *Physica B-Condensed Matter* **397** 179–81
[4] Chupp T E, Coulter K P, Kandes M, Sharma M, Smith T B, Jones G, Chen W C, Gentile T R, Rich D R, Lauss B, Gericke M T, Gillis R C, Page S A, Bowman J D, Penttila S I, Wilburn W S, Dabaghyan M, Hersman F W and Mason M 2007 *Nuclear Instruments and Methods A* **574** 500–9
[5] Petoukhov A K, Andersen K H, Jullien D, Babcock E, Chastagnier J, Chung R, Humblot H, Lelievre-Berna E, Tasset F, Radu F, Wolff M and Zabel H 2006 *Physica B-Condensed Matter* **385-86** 1146–8
[6] Jones G L, Dias F, Collett B, Chen W C, Gentile T R, Piccoli P M B, Miller M E, Schultz A J, Yan H, Tong X, Snow W M, Lee W T, Hoffmann C and Thomson J 2006 *Physica B-Condensed Matter* **385-86** 1131–3
[7] Walker T G and Happer W 1997 *Rev Mod Phys* **69** 629–42
[8] Nacher P J and Leduc M 1985 *Jounal de Physique* **46** 2057–73
[9] Andersen K H, Jullien D, Petoukhov A K, Mouveau P, Bordenave F, Thomas F and Babcock E 2009 *Physica B-Condensed Matter* **404** 2652–54
[10] Parnell S R, Babcock E, Nnighoff K, Skoda M W A, Boag S, Masalovich S, Chen W C, Georgii R, Wild J M and Frost C D 2009 *Nuclear Instruments and Methods A* **598** 774–8
[11] Babcock E, Mattauch S and Ioffe A 2011 *Nuclear Instruments and Methods A* **625** 43–6

[12] Gentile T R, Babcock E, Borchers J A, Chen W C, Hussey D, Jones G L, Lee W T, Majkzrak C F, O'Donovan K V, Snow W M, Tong X, Velthuis S G E T, Walker T G and Yan H 2005 *Physica B-Condensed Matter* **356** 96–102

[13] Tasset F and Ressouche E 1995 *Nuclear Instruments and Methods A* **359** 537–41

[14] Andersen K H, Cubitt R, Humblot H, Jullien D, Petoukhov A, Tasset F, Schanzer C, Shah V R and Wildes A R 2006 *Physica B-Condensed Matter* **385-386** 1134–7

[15] Krycka K L, Booth R, Borchers J A, Chen W C, Conlon C, Gentile T R, Hogg C, Ijiri Y, Laver M, Maranville B B, Majetich S A, Rhyne J J and Watson S M 2009 *Physica B-Condensed Matter* **404** 2561–4

[16] Ioffe A, Babcock E, Pipich V and Radulescu A 2011 *IOP journal of physics conference series, this volume*

[17] Mattauch S, Ruecker U, Korolkov D, Bussmann K, Fracassi F, Schmitz R, Daemen J, Budwig A, Kaemmerling P, Wagener M, Suxdorf F, Kleines H, Bednarek M, Paulzen M, Sparbrodt A, Babcock E, Ioffe A, Richter D and Brckcl T 2008 *JCNS Experimental Reports 2007/2008* (Forschungszentrum Juelich GmbH) pp 52–6 www.jcns.info/wns_reports/

[18] Lauter V, Ambaye H, Goyette R, Lee W T H and Parizzi A 2009 *Physica B-Condensed Matter* **404** 2543–6

[19] Boag S, Babcock E, Andersen K H, Becker M, Charlton T R, Chen W C, Dalgliesh R M, Elmore S D, Frost C D, Gentile T R, Anton R L, Parnell S R, Petoukhov A K, Skoda M W A and Soldner T 2009 *Physica B-Condensed Matter* **404** 2659–62

[20] Chen W C, Gentile T R, Fu C B, Watson S, Jones G L, McIver J W and Rich D R 2011 *IOP journal of physics conference series, this volume*

[21] McKetterick T J, Boag S, Stewart J R, Frost C D, Skoda M W A, Parnell S R and Babcock E 2010 *submitted to Physica B-Condensed Matter*(Available online 3 December 2010) ISSN 0921-4526, DOI: 10.1016/j.physb.2010.11.088.

[22] Gentile T R, Jones G L, Thompson A K, Barker J, Glinka C J, Hammouda B and Lynn J W 2000 *J. of App. Crystallography* **33** 771–4

[23] Schweika W and Boni P 2001 *Physica B Condensed Matter* **297** 155–9

[24] Voigt J, Babcock E and Bruckel T 2010 *Journal of Physics Conference Series* **211** 012032

[25] Stewart J R, Andersen K H, Babcock E, Frost C D, Hiess A, Jullien D, Stride J A, Barthelemy J F, Marchal F, Murani A P, Mutka H and Schober H 2006 *Physica B Condensed Matter* **385-86** 1142–5

Journal of Physics: Conference Series **294** (2011) 012006 doi:10.1088/1742-6596/294/1/012006

Realization of administration unit for ^{3}He with gas recycling

M Güldner[1*], S Becker[2], A Friesenecker[2], K K Gast[3], W Heil[1], S Karpuk[1], E W Otten[1], J Rivoire[3], Z Salhi[1], A Scholz[4], L M Schreiber[3], M Terekhov[3], P Weiss[2], U Wolf[3] and J Zentel[2]

[1]Joh. Gutenberg-University of Mainz, Institute of Physics, 55128 Mainz, Germany
[2]ic-automation GmbH, 55130 Mainz, Germany
[3]Joh. Gutenberg-University of Mainz, Department of Radiology, 55128 Mainz, Germany
[4]Joh. Gutenberg-University of Mainz, Department of Anesthesiology, 55128 Mainz, Germany

E-mail: gueldner@uni-mainz.de

Abstract. Hyperpolarized (HP) noble gases (^{3}He,^{129}Xe) are used for MR-imaging of the lung. In the majority of case the HP gas is filled in Tedlarbags and directly inhaled by the patients. Starting from an earlier pilot device, an administration unit was built respectively to the Medical Devices Law to administer patients HP noble gas boli in defined quantities and at a predefined time during inspiration with high reproducibility and reliability without reducing MR-quality. The patient's airflows are monitored and recorded. It is possible to use gas admixtures, measure the polarization on-line and collect the exhaled gas for later recycling. The first images with healthy volunteers were taken with this setup in a clinical study. Current results will be presented.

1. Introduction

Hyperpolarized (HP) noble gases are used for MR-imaging of the lung for many years[1]. In case of the rare isotope ^{3}He this technique requires efficient polarization, storage, transport, gas administration and recovery of the gas [2, 3]. In recent development all steps of this cycle have been optimized. In the beginning the HP gas was filled in Tedlarbags and directly inhaled by the patients. However for follow-up studies a better reproducibility and monitoring is desirable. Starting from an earlier pilot instrument [5, 6] which has been used in the European PHIL-study [7], an advanced administration unit was built which will be presented in the following.

2. Requirements of the administration unit

Tedlarbags (Fig. 1) were used in ^{3}He-MR-lung-Imaging for many years. Because of this experience several problems with the use of Tedlarbags are well known:
First, automatic filling is not available; so the reproducibility of the bolus volume strongly depends on the user. Spirometry is not included; therefore no information about the breathing parameters as the depth of the inhalation and the flow rate is available. This means the amount of gas remaining inside the Tedlarbag after inhalation is unknown, too. Further, Tedlarbags are only available with a very thin tube (inner diameter about 3 mm). So the patients have to inhale against an unphysiological high breathing resistance. This effect can be seen in the dynamic images which were taken in the clinical study presented below. Gas collecting for later recovery is difficult, as the gas is inhaled via the thin tube with the mouth, but for gas collecting a mask is needed. And without using a pump Tedlarbags can be filled only with one bar remaining in the gas cell.

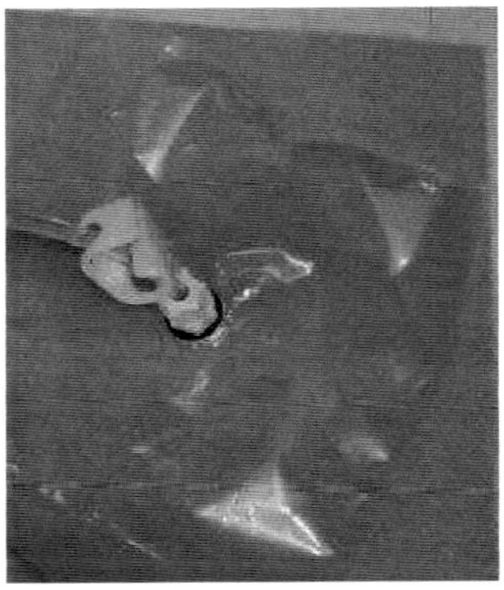

Figure 1. Photograph of a Tedlarbag.

So these problems form the requirements for the administration unit: First of all the filling should be automatic and controlled via laptop. There should be a high reproducibility of the preset bolus volume (accuracy: $\Delta V / V < 3\%$) and the time of administration (accuracy: $\Delta t < 100$ ms); with the administration unit it is also possible to predefine the time during inhalation when the gas bolus should be administered, for example either at the beginning or 300 ms after the start of the inhalation. The presented administration unit can further be equipped with an on-line polarimeter to monitor the HP gas polarization with an accuracy of $\Delta P / P < 7\%$. It is essential to conserve the polarization during the whole process (T_1 has to be as long as possible), which constricts in addition to the required biocompatibility, the choice of materials used for the construction of the administration unit. The control and monitoring should be as easy and comfortable as possible via laptop, especially the breathing curve should be visible all the time. The automatic filling should be user-friendly such, that one can save typical administration settings and the log-files should be stored automatically. Especially for structure analysis like measuring the apparent diffusion coefficient (the ADC), it is useful to have an option to administer gas admixtures, as different gases such as air, N_2, SF_6 or Xe have an influence on the apparent diffusion coefficient. As ^{3}He shortage is a serious problem efficient gas collecting for later recycling is included in the administration unit presented below.

One of the most important aspect for the construction of the presented administration unit concerns the obligations (safety and hygienic aspects) given by the Medical Devices Law.

3. Realization of the ^{3}He administration unit

In the picture of the setup Fig. 2 on the left compensation coils (blue) provide a homogenous field at the HP gas cell position in the stray field of the tomograph. The sketch of the setup (Fig. 3) point the magnetic field lines up. The piston pump which controls the administered HP gas bolus is placed below. Behind the patient's head the administration box is positioned containing spirometers and valves (Fig. 2). The patient is wearing a breathing mask and is connected directly to the administration box. On the floor the bag collecting the exhaled air can be seen. Later on the exhaled air/^{3}He mixture is compressed into a steel bottle and shipped to a central recycling station which recovers the ^{3}He with 95% efficiency and ppm purity level [3, 4]. Around the patient's thorax one recognizes the RF-coil to excite and receive the MR-signal. Outside the scanner room the control unit is placed. In the following all components will be described in detail.

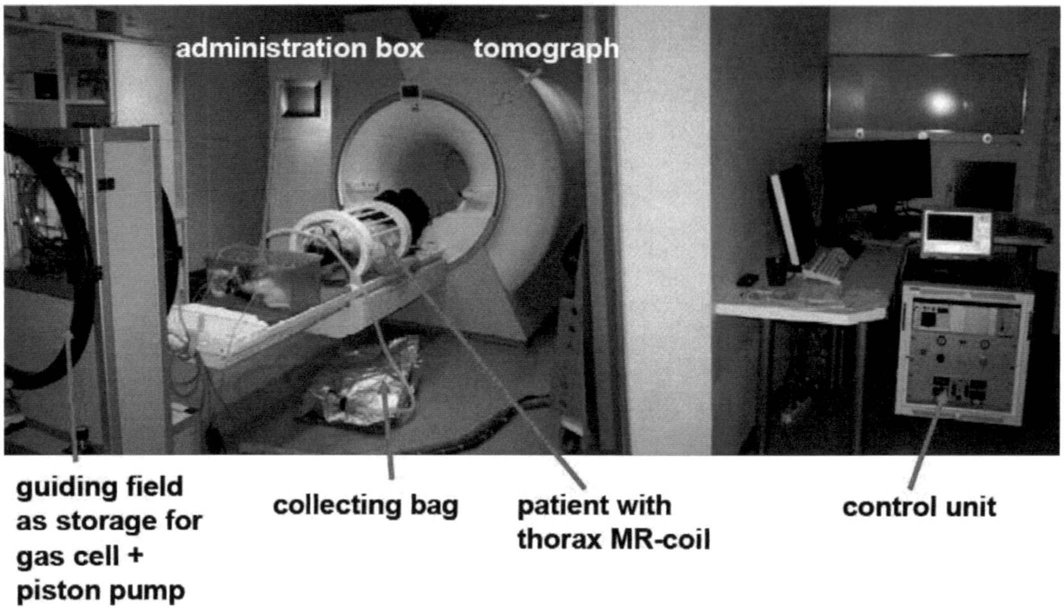

Figure 2. Photograph of the ^{3}He-administration unit setup.

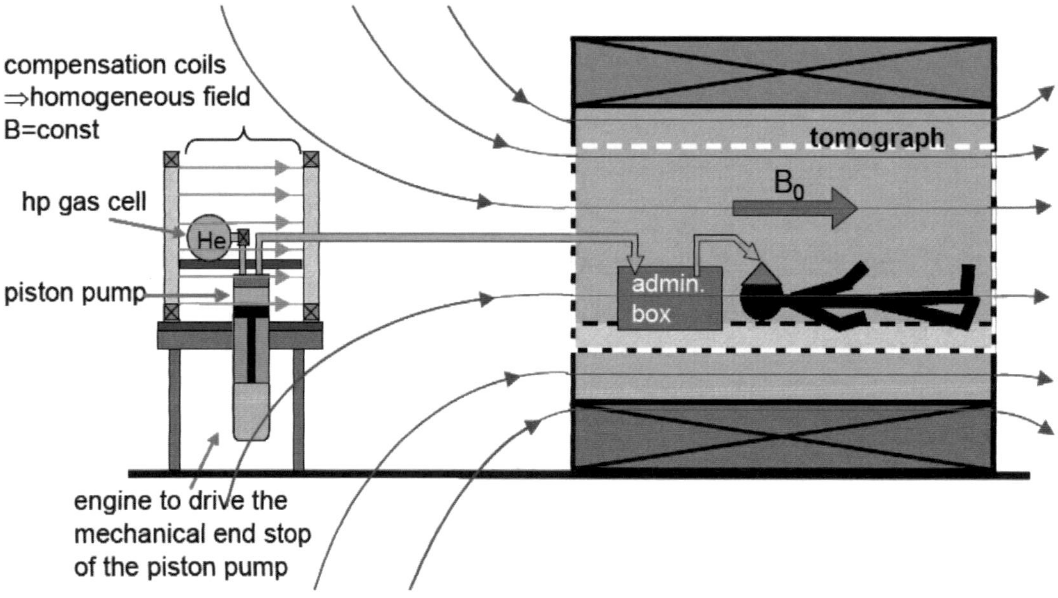

Figure 3. Sketch of the ^{3}He-administration unit setup.

4. Gas storage and filling unit

A pair of coils with a diameter of 0.7 m compensates the stray field of the scanner. The coils have 150 windings each; the distance between the two coils is 0.35 m. In contrast to the Helmholtz-configuration the current in each coil can be regulated separately. The height of the table can be changed, depending of the given height of the scanner. Thus, the administration unit can used at different scanners without changing the construction. For example at the distance of 3 m from the Magnetom Siemens Sonata scanner in the Universitätsmedizin Mainz, where the clinical study was accomplished, the height was 1.01 m, a current of + 0.5 A for the first coil and -0.5 A in the second one was calculated in a simulation and set at the power-supply-unit of the control-unit to compensate the field

Journal of Physics: Conference Series **294** (2011) 012006 doi:10.1088/1742-6596/294/1/012006

gradient [8]. (For this calculation the stray field of the scanner was measured once without compensation and different configurations were then simulated to achieve the best compensation. For each scanner this optimization has to be done separately, as the stray fields vary. In this case the strength of the stray field at the position of the storage cell varied from 3.5 to $4.8 \cdot 10^{-4}$ T without compensation (Fig. 4). Using the compensation coils the field strength is about $3.8 \cdot 10^{-4}$ T at the storage cell position and the field homogeneity over the cell volume is $3 \cdot 10^{-6}$ T/cm. The spherical HP storage gas cell has a volume of about 1.14 l and is filled up to a pressure of 2.7 bar, (the upper limit is given by the IATA Dangerous Goods Regulations).

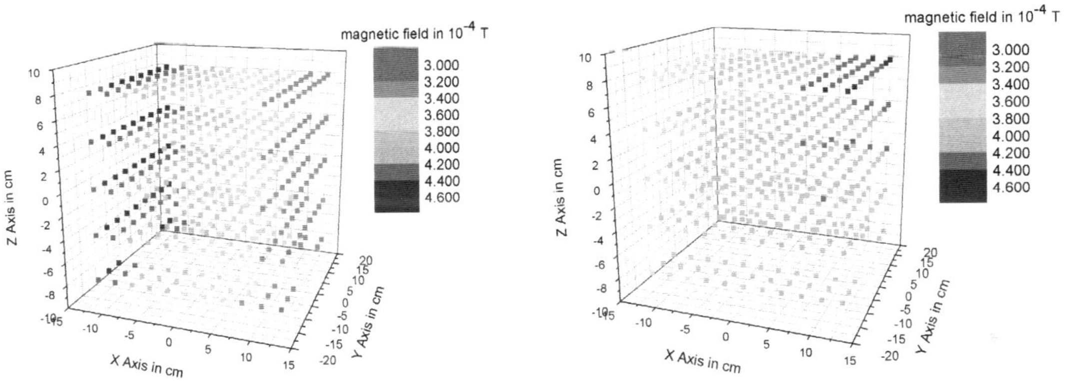

Figure 4. Magnetic field strength in the region of the storage cell volume without (left) and with (right) compensation coils.

Inside thus homogenized field the HP storage gas cell and the piston pump, which defines the amount of administered gas, are placed. More over, it allows using up the HP gases from the storage cells down to a residual pressure of about 200 mbar. The piston as well as the cylinder head, shown in Fig. 5, is manufactured of titanium to minimize the relaxation of the gas. A buffer volume behind the compression volume is added, which is flushed with N_2 to avoid contact to the oxygen of ambient air. To regulate the pneumatically driven stroke of the piston and thus the amount of gas sucked into the compression volume and administered to the patient, a mechanical end stop is mounted at the piston pump. It is driven by a non-magnetic attraction by the scanner). pneumatic engine (to avoid magnetic First the physician has to define a bolus volume. At the control unit typical settings can be stored in and loaded from setting files.

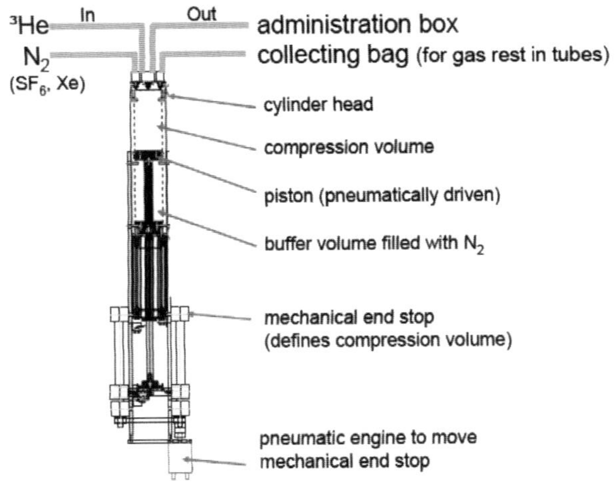

Figure 5. Sketch of the piston pump.

In the presented clinical study 0.2 standard litre (stdl) ^{3}He were used and 0.3 stdl of N_2 were added, except for measurements of the partial oxygen pressure (pO_2) in the lung, where 0.2 stdl pure ^{3}He were used. The used hp gas polarization for all volunteers was 60-65% (the MR-signal intensity is direct proportional to the hp gas polarization). The volume of administered hp gas was kept constant and independent of this small variation of polarization. Before administration the piston is moved into top position and the pressure in the gas cell is measured. Then the control unit calculates the required piston volume depending on the pressure in the gas cell. In the next step the mechanical end stop is driven into the calculated position. Finally the piston transports the gas to the administration box via a thin tube (to minimize the dead volume). As the gas transfer is quite fast (about 0.3 l/s) shielding is not required. The material of this tube, S-50-HL of the company liquidscan, is Tygon, the inner diameter is 2.4 mm with a length of 3 m.

5. Administration box

The administration box is positioned behind the patient's head. Fig. 6 shows the sketch of the administration unit, containing spirometers and valves. The patient is wearing a breathing mask with a microbiological filter/HME. This minimizes the humidity and fulfills hygienic obligations. The patient valve is a membrane valve and it is added to strictly separate inhalation from exhalation via the pressure on the two membranes.

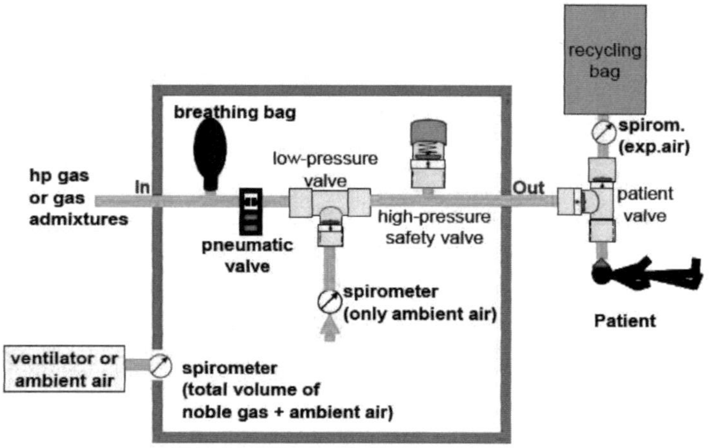

Figure 6. Diagram of the administration box.

The inhalation tract is directly connected to the outlet of the administration box. The patient must not be directly connected to the HP gas or gas admixture arriving from the piston pump on the left side (see Fig. 6). This would force a barotrauma to the patient's lung because the pressure might considerably exceed the atmospheric pressure. Therefore safety provisions are added inside the administration box. To disconnect the patient from any excess pressure, the gas is filled into a storage volume (breathing bag), from where the patient can inhale the gas at atmospheric pressure. The relaxation times T_1 from many breathing bags were measured: Therefore different bags were filled with 0.5 l pure ^{3}He each at atmospheric pressure and every minute a FID in a NMR-system of the institute of physics was started. The best results (with T_1 about 10 min in this setup) were achieved with the breathing bag of chloroprene manufactured by the company Draeger (http://www.draeger.com/media/10/01/80/10018034/accessories_catalog_compl_9066485_en.pdf).

The administration is controlled via a pneumatic valve which is positioned directly behind the breathing bag. This valve is permanently closed, in particular during filling of the breathing bag. It is only opened for administration of the HP gas. To ensure that there is no accidental high pressure towards the patient's lung (in case of a failure of the administration valve) an additional safety high-pressure valve was added. It is a membrane valve, with the membrane being pressed by a non-magnetic spring. This security valve is positioned directly in front of the outlet of the administration box. It obviates pressures higher than 20 mbar above atmospheric pressure.

When the storage breathing bag is filled and the administration valve is open the patient first inhales the amount of gas stored inside the breathing bag. When the content of the breathing bag is completely inhaled, an additional opening is needed which allows the patient to continue breathing. This is achieved with a membrane valve, too. If the bag is empty the resistance on the patient's side of the membrane valve is increasing and the patient is then inhaling ambient air through the membrane then. To monitor the amount of inhaled ambient air (flow rate), a spirometer was added. As the air has to flow into the box an opening in the box is needed. A ventilator can be connected there as an option or the ambient air just flows into the box. To measure the total volume of inhalation another spirometer is mounted in this place as well. From the difference between both flow signals the control unit calculates the HP gas-flow rate and thus the administered volume as integral of the flow of HP gas.

For monitoring the expiration air flow a third spirometer is added. To collect the gas for later recovery a collecting bag is used in which the patient exhales. Details about the used collecting bag are given below. In the presented clinical study 25 breathing cycles after administration have been collected. For a complete washout of ^{3}He from the lung 20-25 consecutive breathing cycles should be collected. Hence this bag should be rather big (100-200 litres); it also has to be reasonably ^{3}He tight. To minimize the volume of the collected gas one can add a three-way-valve in front of the collecting bag and switch it towards ambient air during ^{3}He-free phases (as it was realized in the presented study). As the three-way-valve is not essential, it is not included in Fig. 6, to gain a better overview.

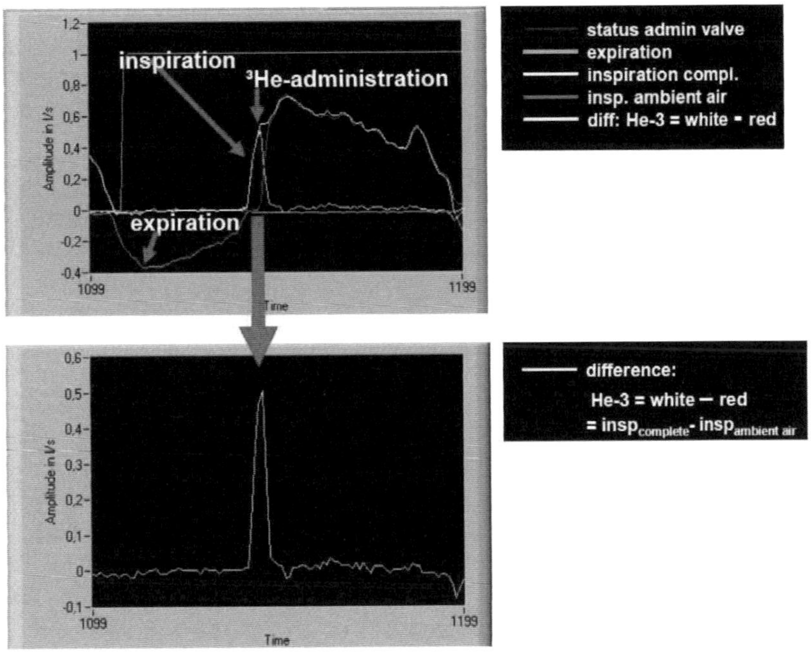

Figure 7. Monitored signals of the spirometers.

6. Monitoring and control unit

The control unit itself is placed outside of the scanner room and it is operated via a laptop computer. The currents of the two compensating coils can regulated there (via the two integrated power-supply-units) according to the calculated value (see Sec. 4), as well as the control of the piston pump and the pneumatic administration valve of the administration unit. The signals of the spirometers are converted and monitored. Fig. 7 shows an example of the monitored signals. The upper curve (white) is showing the complete inspiration, while the lowest curve (green) the expiration. The red line is showing the signal of the flow of ambient air. When the HP gas bolus is administered a delayed increase is observed. This delay is due to the fact that the patient is inhaling the HP gas admixture first until the bag is empty. After that he starts to inhale ambient air, which causes the increase of this flow rate. If one does not want to administer the HP gas admixture at the beginning of the inhalation this shape is shifted to the right. From the difference of both signals, which is shown in yellow, the control unit calculates the flow of the HP gas admixture. In the beginning of the construction the cables of the spirometers have generated MR-noise. After accurate cable screening the whole administration unit does not increase the noise level and thus does not reduce the MR-quality as proved by Fig. 8.

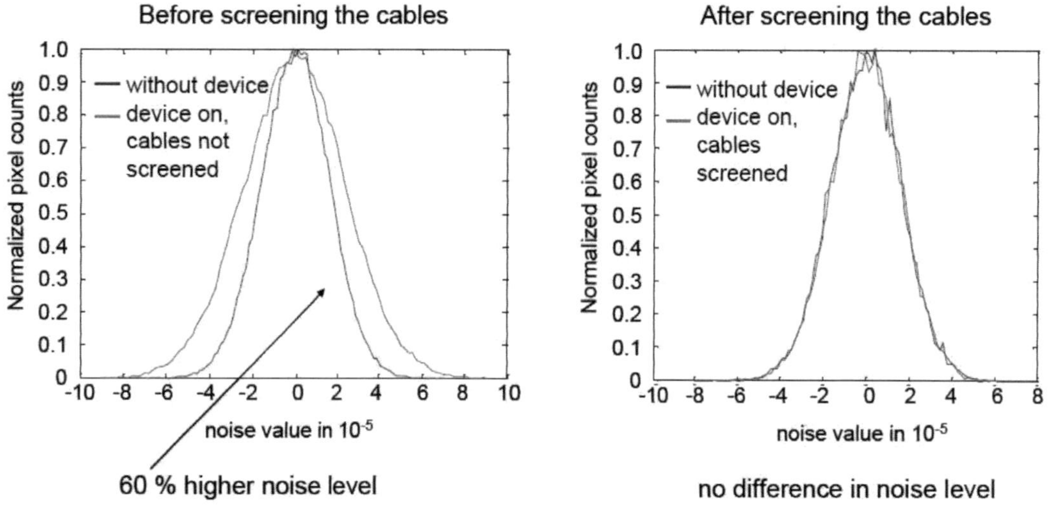

Figure 8. Noise level.

7. Clinical study

First images of human lungs (healthy volunteers) were taken with this setup in a clinical study. These study was performed in the Department of Radiology at the Johannes-Gutenberg University Mainz using the Scanner "Sonata" Magnetom (Siemens Medical Solution, Erlangen) providing a magnetic field of 1.5 T. As RF-coils a double-tuned elliptical ^{3}He-^{19}F TX/RX birdcage resonator (Rapid Biomedical) and a Spoiled Gradient Echo (SGRE) pulse sequence was used.

^{3}He-lung-MRI features four typical methods [1, 6, 7, 9-15]: a) morphological imaging at apnoea, which gives the spin density, b) measurement of the partial oxygen pressure (pO2), which makes use of the T_1-relaxation of HP ^{3}He by the paramagnetic oxygen, c) measurement of the apparent diffusion coefficient (ADC), which depicts the fine structure of the lung as this structure restricts diffusion and d) "dynamic ventilation" imaging, which visualize the process of ^{3}He distribution in lung during in- and exhalation with high temporal resolution (10 fps). From these methods the following parameters are available: from morphological images the ventilation of the lung and the signal-to-noise-ratio (SNR), from b) and c) the parameters pO2 and ADC, from dynamic imaging the parameter "rise-time", which characterize the rate and amount of ^{3}He delivery in different lung regions (influenced by the inflow of HP gas as well as by its relaxation in the lungs).

The clinical study compares the reproducibility of these MR-parameters of images taken by the administration unit with those taken by a Tedlarbag. To test the reproducibility, images were taken using each method twice, with Tedlarbag and as well as with the

Journal of Physics: Conference Series **294** (2011) 012006 doi:10.1088/1742-6596/294/1/012006

administration unit. This means 16 measurements on each of the 10 healthy male volunteers which were all non-smokers. These data will be analyzed once for each volunteer separately to check the intra-individual reproducibility and once for the whole group of volunteers checking the inter-individual reproducibility. The analysis of this data is in progress.

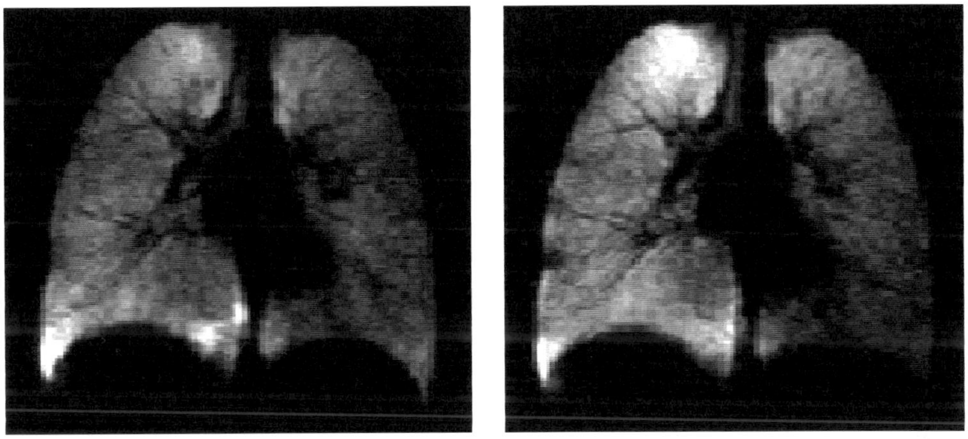

Figure 9. Morphologic Images taken by administration unit (left) and Tedlarbag (right).

Fig. 9 shows a morphological image taken by the administration unit in comparison to an image taken by Tedlarbag. The image quality is in a comparable range except that the right image is a bit brighter. The used matrix size was 128x128, the field of View (FOV) 320 mm x 320 mm, the slice thickness 10 mm, the flip angle 18°, the repetition time (TR) 11 ms and the echo time (TE) 4 ms. The SNR was analyzed according to the method described in [16]. The SNR was calculated using the software ImageJ (version 1.43u; National Institutes of Health, Bethesda, MD, USA). The mean signal intensity (SI) was measured in six circular (diameter = 27.5 mm) homogeneous regions of interest (ROIs) within the lung on each MR image (SI_{uncorr}). In analogy the noise signal intensity (SI_{noise}) was measured in six circular (diameter = 27.5 mm) homogeneous regions of interest (ROIs), too. These ROIs were placed at the corresponding height in the morphological image but outside of the lung. Noise correction was performed using the approach described by Gudbjartsson and Patz [17].

While the highest values of SNR of images taken by Tedlarbag are bigger than those images taken by the administration unit, the SNR-values of the administration unit scatter less, see Fig. 10. The reproducibility still has to be calculated. In all images the signal of ROIs in the right lung of this particular volunteer are higher than of those in the left lung. This effect can be seen in Fig. 9, too, where the signal of the patient's right lung (left on image) is enhanced. It is a hardware effect caused by the proximity of the coil and the gradient system (which was established by several tests).

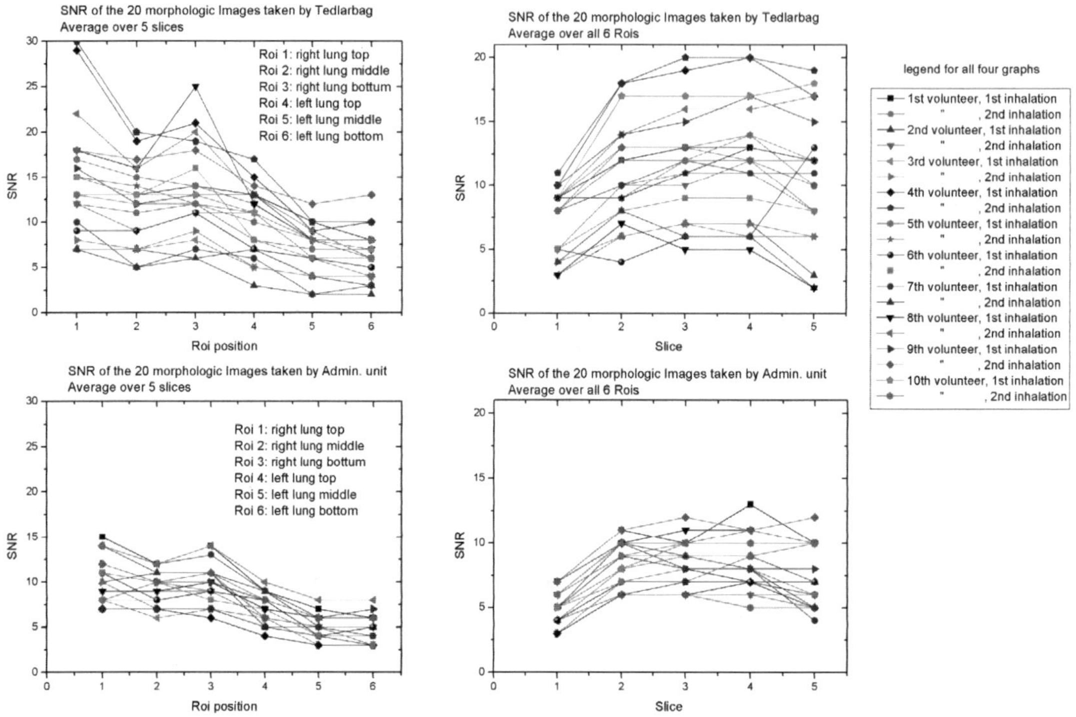

Figure 10. SNR of morphologic images taken by administration unit below and Tedlarbag above.

For dynamic imaging the administration unit clearly shows an advantage: The administration unit provides a very homogeneous intake of the gas. It is not yet clear if it is homogeneous in time or homogeneous in space. Looking at a movie taken by the Tedlarbag, on the other hand, one clearly observes the perturbation of signal-time curve during the time when the volunteer finishes with inhaling the ^{3}He from Tedlarbag and after short breath hold changes to breathing in of the ambient air. In the movie it seems like a small exhalation. There is some sort of stock in the flow, the signal decreases a little and increases again. This effect can also be recognized in the graph of the MR-signal (Fig. 11). In the images taken by the administration unit one observes a steady rise of the signal during inhalation of HP gas followed by a steady decrease caused by the combined effects of exhalation and relaxation of the HP-gas. The detailed analysis of all MR-parameters and physiologic interpretation is still in progress.

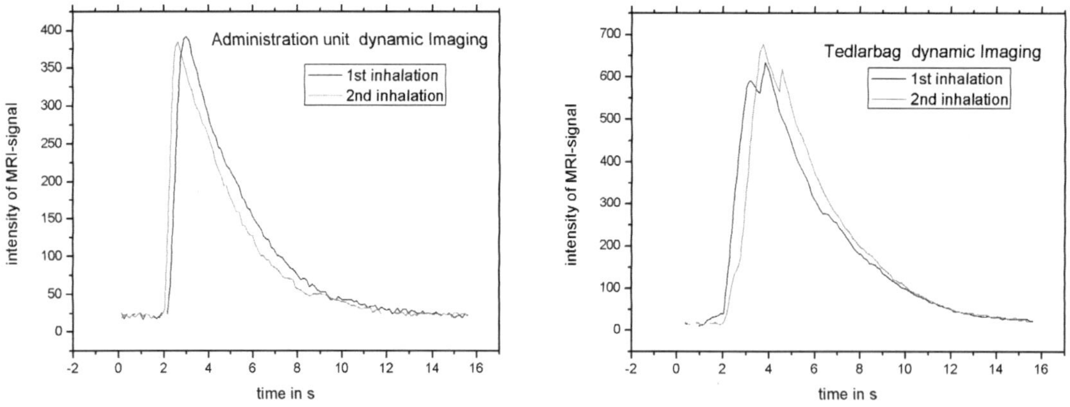

Figure 11. Dynamic Imaging: Intensity of MR-signals in images taken with Administration unit (left) and Tedlarbag (right).

Journal of Physics: Conference Series **294** (2011) 012006 doi:10.1088/1742-6596/294/1/012006

8. Collecting and recovery [3, 4]

In the presented study 25 exhalations after HP gas administration have been collected in a ^{3}He-tight bag "Tecobag" manufactured by the company Tesseraux. The maximum capacity is 150 litres. The inner slice is made of Polyethylen, the middle slice is Aluminium, covered with Polyethylenterephthalat. A three-way-valve was used to switch between collecting exhaled air in the Tecobag and exhalation into the ambient air. When the collecting bag is full, its content is compressed into a steel bottle. The respective compressor, MCH6/EM manufactured by the company Aerotecnica Coltri, shown in Fig. 12, was designed for filling diving bottles. The collecting efficiency achieved in this clinical study was 80-83 %.

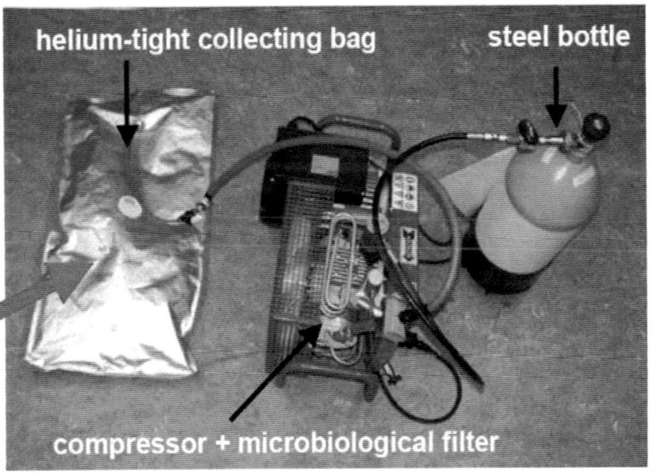

Figure 12. Gas collecting in the scanner room (left), compressing collected gas into steel bottle (right)

After collecting and compressing the exhaled gas, it was transferred into a recovery unit at the institute of physics of the university Mainz. The percentage of He inside the mixture is usually quite low (around 1%), but the removal of all contaminant gases by LN_2 cooled zeolith traps, followed by a cryopump working at 10 K is working very efficiently. Thus the recovery efficiency of this unit is 95% and the gas purity is in the region of ppm [3, 4]. The excellent gas purity has been confirmed by repolarizing it in our polarizer (Fig. 13). The graphic shows almost no difference between polarization curves obtained with fresh gas from the gas supplier and the recovered one. The whole recycling process is included now in our Manufacturing Authorization and allows re-using the recovered gas for clinical studies on humans.

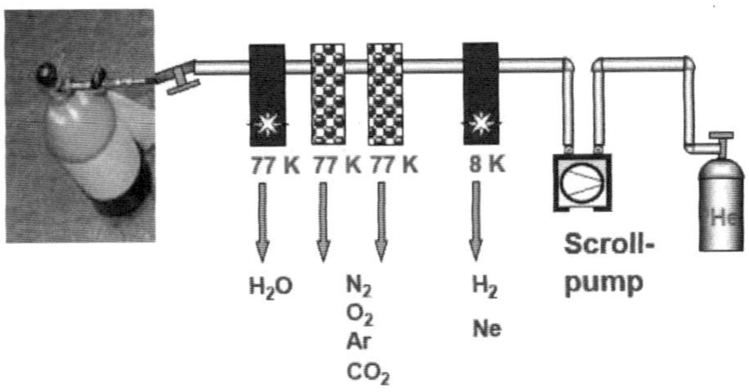

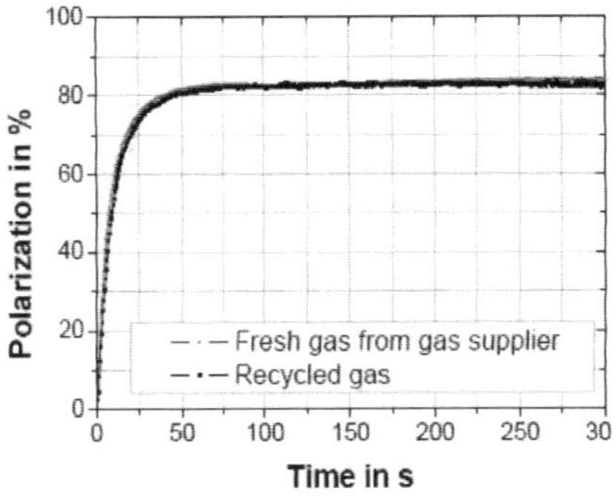

Figure 13. Recovery unit (left), polarization curve (right).

The total efficiency at the presented study was 76-79% as a result of the multiplication of the collecting and the recovery efficiency. The very limited stock and production rate of ^{3}He on one side and the growing demand on the other have led to a dramatic shortage and prize explosion in recent years making gas recovery mandatory. In this study a total of 32 stdl HP ^{3}He have been administered of which about 78 % have been recovered resulting in a loss < 10 stdl. Already modest increase in the total efficiency would cause a big reduction in loss of usable ^{3}He. The aim is to increase the efficiency to 85% by optimizing the collection efficiency of the exhaled gas.

9. Summary

The presented administration unit allows fully automated administration of HP gases with high accuracy in volume and time. It shows great advantages for dynamic imaging and allows use of gas admixtures. Further it fulfills the obligations given by the Medical Devices Law and enables efficient and easy gas collecting for later recycling. The administration will now be tested on patients and further research is planned to increase the efficiency of gas collecting.

Acknowledgement

This project is supported by the BMBF (01EZ 0612), the "Kompetenznetzwerk COPD/Asthma", the DFG (FOR474) and the Carl-Zeiss-Stiftung.

References

[1] Möller H E et al 2002 *Magn. Reson. Med.* **47** 1029-51
[2] Schmiedeskamp J 2004 Ph.D. thesis of Johannes Gutenberg-Universität Mainz
[3] Großmann T 2000 Ph.D. thesis of Johannes Gutenberg-Universität Mainz
[4] Salhi Z et al to be published
[5] Lauer L 1997 diploma thesis of Johannes Gutenberg-Universität Mainz
[6] Eberle B 1999 *J. Appl. Physiol.* **87** 2043-52
[7] Bletz C 2002 *Fortschr. Roentgenstr.* **83** 263
[8] Glück F, KIT Karlsruhe to be published.
[9] Deninger A J et al 1999 *J. Magn. Reson.* **141** 207–216
[10] Deninger A J 2000 Ph.D. thesis of Johannes Gutenberg-Universität Mainz
[11] Guenther D et al 2000 *NMR Biomed.* **13**
[12] Schreiber W G et al 2000 *Fortschr. Röntgenstr.* **172** 129–133
[13] Gast K K et al 2008 *Invest. Radiol.* **43** 439–46
[14] Morbach A 2006 Ph.D. thesis of Johannes Gutenberg-Universität Mainz
[15] Kauczor H-U 2000 Franz Steiner Verlag 1st Edition
[16] Wolf U et al. 2006 *Magn. Reson. Med.* **55** 948–51
[17] Gudbjartsson H, Patz S 1995 *Magn. Reson. Med.* **34** 910–914

Journal of Physics: Conference Series **294** (2011) 012007 doi:10.1088/1742-6596/294/1/012007

Construction of a compact ^{3}He polarizing facility

C Mrozik*, O Endner, C Hauke, W Heil, S Karpuk, J Klemmer and E W Otten

Institut für Physik, Johannes Gutenberg-Universität Mainz, Staudinger Weg 7, 55128 Mainz, Germany

E-mail: christian.mrozik@uni-mainz.de

Abstract. In this paper we present concepts, developed to construct a compact ^{3}He polarizing facility as well as first results of their realization. The apparatus, which is currently in the state of construction, is based on the method of metastability exchange optical pumping (MEOP). Contrary to the present apparatus at the university of Mainz, which serves as central polarizing facility, the compact polarizer is designed to serve as local polarizing facility in both basic research and medical application. With the new polarizer we aim to reach polarization degrees of $P > 65\%$ at a flux of several standard liters per hour.

1. Introduction

Applications of hyperpolarized ^{3}He are various. The strongly spin-dependent absorption of neutrons in polarized ^{3}He opens up the possibility of polarizing neutrons over the full kinematical range of cold, thermal and hot neutrons [1]. Hyperpolarized ^{3}He nuclei are as well used as kind of polarized neutron target in medium energy physics [2] and in precision magnetometry [3]. In 1994 it was discovered, that hyperpolarized gases can serve as contrast agent in medical imaging in order to display lungs in vivo [4]. The wide interest in this new medical diagnostic tool made it necessary to produce large quantities of hyperpolarized ^{3}He. Therefore a polarizing facility has been developed at the university of Mainz a decade ago, designed for centralized polarization of ^{3}He. The concept of remote production of hyperpolarized ^{3}He requires means for shipping the polarized gas safely to the customer. To that end glass vessels have been blown from a special melt of He-diffusion tight glass, showing little wall relaxation [5]. For storage and transport they are packed in specially developed spin boxes with a homogeneous magnetic field inside, that maintains the polarization of the gas [6]. The residual relaxation rate is typically about 1/(100 h).

The design of the compact polarizing facility currently under construction facilitates local polarization. As it will be possible to install the polarizer directly at an experiment in order to perform *on-line polarization*, experiments that require the highest possible degree of polarization can benefit from the compact facility. On the other hand, users needing huge amounts of polarized ^{3}He e.g. for clinical applications can gain independence from gas-deliveries, when using the compact polarizer at their own facility.

2. The current polarizing facility

The current polarizing facility is based on the method of metastability exchange optical pumping (MEOP) [7]. ^{3}He gas at a pressure of about 1 mbar is polarized in a weak gas discharge by two

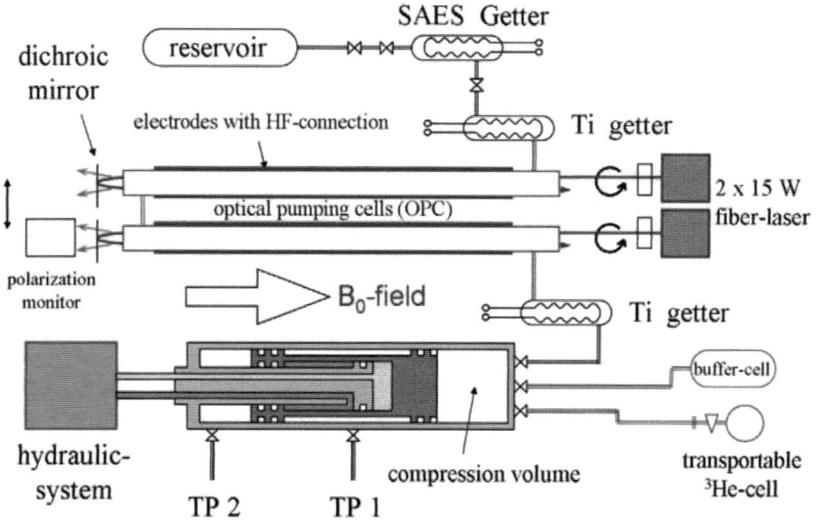

Figure 1. Sketch of the current MEOP polarizing facility at Mainz. It is located in a homogeneous magnetic field of 1 mT provided by a set of large air coils which occupy a volume of 2.6 m in length and 1.6 m in diameter. Shown are two of the five optical pumping cells with a total volume of 36 l. [8][9][10]

fiber-lasers at 1083 nm. The whole apparatus is located in a homogeneous magnetic field of 1 mT (relative field gradients $(dB_r/dr)/B_0$ in the order of $10^{-4}\,\mathrm{cm}^{-1}$), which serves as quantization axis and holding field for the ^{3}He-spins. The polarizer consists of three parts.

The first part (upper part in fig. 1) consists of the reservoir for the unpolarized gas and getters for gas purification. These are necessary as gas impurities exceeding the ppm level reduce the density of metastable atoms in the gas discharge and hence the degree of polarization, that can be reached in the optical pumping.

The second part (middle part in fig. 1) contains the optical pumping volume with typical pressure values of about 1 mbar. A weak gas discharge excites Helium atoms into the metastable 2^3S_1 ground state. Optical pumping is performed with two 15 W fiber lasers at 1083 nm. After having passed a polarizer cube and a $\lambda/4$ plate the light is circular polarized and gets absorbed by the metastable atoms. In this absorption process angular momentum of the photons is transferred to the electron shell of the atom. By hyperfine-coupling the angular momentum gets transferred to the nucleus. Via metastability exchange collisions the ^{3}He atoms get back to their ground state.

The third part (lower part in fig. 1) contains a mechanical, polarization-conserving piston compressor driven by hydraulics in order to achieve gas pressures up to 5-6 bar. In a first step the gas is compressed into a buffer cell of $V = 6\,\mathrm{l}$. After having polarized the desired amount, the polarized gas from the buffer cell is compressed in a second step into a detachable transport cell.

Fig.s 2 and 3 show the performance of the current polarizer. The build up of the nuclear polarization is shown in fig. 2. In steady state a maximum polarization of (84±2)% can be achieved with the facility. Fig. 3 shows the polarization of the gas in dependency from the throughput of hyperpolarized ^{3}He. For medical applications, where high production rate counts more than ultimate polarization, a working point around 65% polarization and 3 standard liters per hour throughput is adequate. Basic research, on the other hand often requires the highest

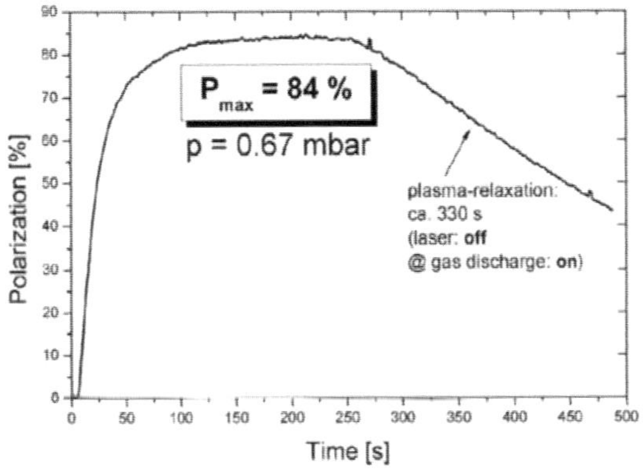

Figure 2. ^{3}He polarization in the current facility as function of time. [1]

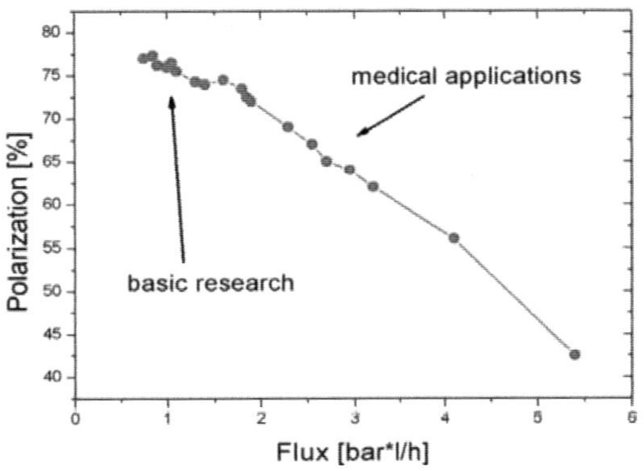

Figure 3. Relation between ^{3}He polarization and throughput of hyperpolarized gas.

possible degree of polarization. In that region the throughput drops to about 1 standard liter per hour.

3. The new compact polarizing facility

The new compact facility follows the same design principles and aims at the same performance data as its predecessor (see fig.s 1 - 3). However, a decisive reduction in size is achieved by embedding the critical components shown in fig. 1 into a magnetic shield with a homogeneous field inside. Moreover, we have tested and will implement a more efficient concept for the various optical components which allows - amongst others - saving of one of the two high power pump lasers. Also the cumbersome hydraulic compressor drive will be replaced by a neat linear motor.

3.1. Magnetic field and shielding

In order to keep the relaxation of the polarization due to a magnetic field gradient small, a sufficiently homogeneous magnetic field has to be provided. T^{grad}, the time constant of relaxation induced by diffusive motion of the gas atoms through an inhomogeneous magnetic field, is given by

$$\frac{1}{T^{grad}} = D \cdot G_r^2$$

D is the diffusion coefficient and G_r the relative transverse field gradient

$$G_r = \sqrt{\left(\vec{\nabla} B_x\right)^2 + \left(\vec{\nabla} B_y\right)^2} / B$$

Using the known diffusion coefficient at room temperature one can write T^{grad} as [6]

$$T^{grad}\,[\mathrm{h}] = \frac{1}{6900} \frac{p\,[\mathrm{bar}]}{G_r^2\,[\mathrm{cm}^{-2}]}$$

With a desired $T^{grad} \geq 1\,\mathrm{h}$ the radial field gradient has to be $G_r < 3.8 \cdot 10^{-4}\,\mathrm{cm}^{-1}$ at 1 mbar. Such a homogeneous magnetic field can be provided by a solenoid. To construct a compact apparatus it is imperative to use the volume of such a solenoid efficiently. The concept of creating a homogeneous field all over the solenoid's volume consists of covering it at both ends with flat polefaces and returning the flux through a mantle around the windings [6] (see fig.

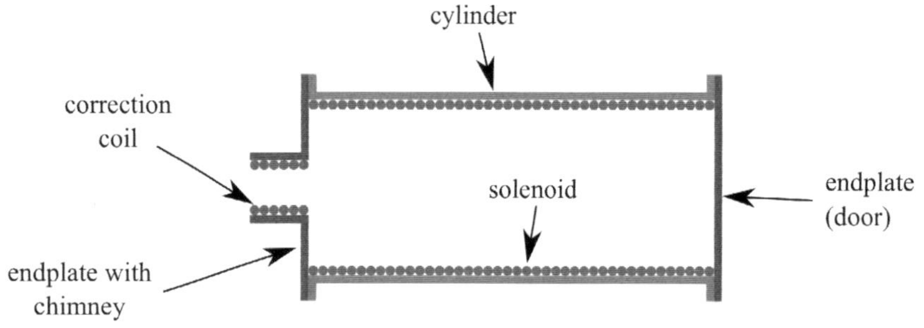

Figure 4. Solenoid enclosed in soft magnetic material.

4). This closed cylinder, built from material with high magnetic permeability [11], provides also an efficient shielding of magnetic stray fields which allows to place any electronics, motors, steel constructions etc. close by. One of the endplates has an opening in order to feed in a support structure from aluminum as well as supply lines for helium, vacuum, pneumatics, electronic signals etc. to the assembly inside the solenoid. As the field homogeneity inside would be disturbed by this opening, a chimney with a correction coil is attached to the opening. The correction coil is continuing the homogeneous field into the chimney. The endplate on the opposite side is designed as a door in order to be able to remove a transport cell, after it has been filled with polarized gas.

Two-dimensional simulations using the fast FEM program *femm 4.0* ©1998-2003 by D. Meeker show, that this design yields the desired field-homogeneity within almost the total 80 cm

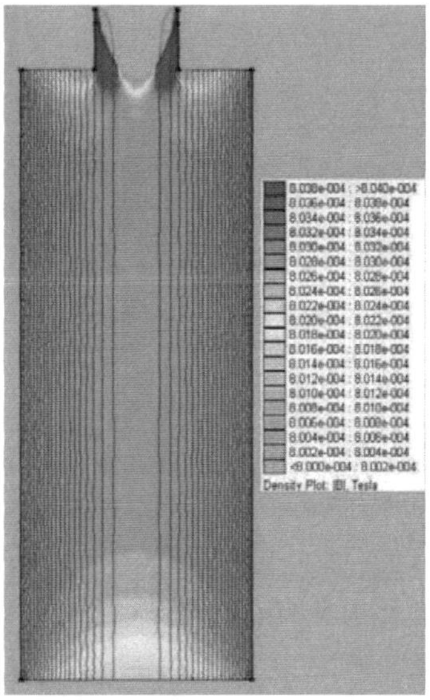

Figure 5. Simulation of the magnetic field inside the shielded solenoid. One colour step comprises a relative field interval of $2.5 \cdot 10^{-4}$.

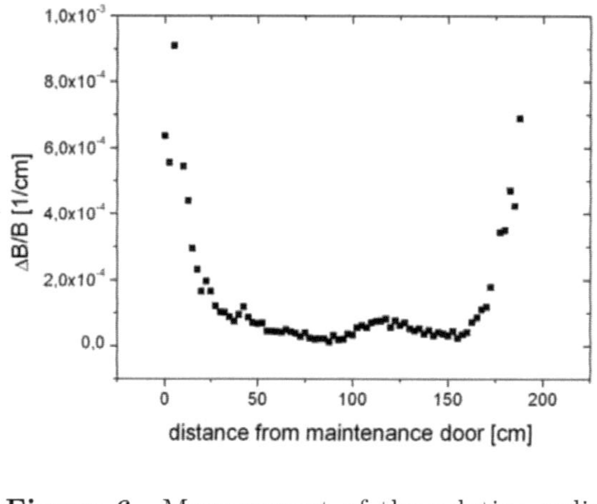

Figure 6. Measurement of the relative radial field gradient. The data is taken on an axis where OPC 2 is located (see fig. 8), starting at the maintenance door and proceeding to the endplate with the chimney.

Journal of Physics: Conference Series **294** (2011) 012007 doi:10.1088/1742-6596/294/1/012007

wide and 200 cm long cylinder (see fig. 5). Fig. 6 displays measured data in the realized object. The measurements show that the required field homogeneity is being reached everywhere in the solenoid but the first 25 cm next to the endplates, which results in a 1.5 m long cylindrical volume of 80 cm in diameter for assembling the decisive components of the polarizer shown in fig. 1. Measurements of the total relaxation time in a long optical pumping cell resulted in $T_1 = 330$ s, which is the same as in the cells of the present polarizer (see fig. 2), where sufficient conditions for polarizing are found.

3.2. New concept for the optics

The optical pumping requires a circularly polarized laser beam at a wavelength of λ=1083 nm. Unlike the current polarizing facility, which uses two fiber lasers, the compact apparatus will be equipped with only one pump laser. ^{3}He gas is optical pumped in six 1.25 m long borosilicate glass tubes with a total volume of about 19 l. The beam from a 15 W fiber laser is magnified with a telescope to a diameter of about 35 mm and then guided through the six glass tubes. Polarizing beam splitters with attached $\lambda/4$-plates are put ahead of each glass cell to restore a perfect circular polarization of the incoming beam and to deflect the returning outgoing beam towards the next cell (see fig. 7). After each cell a dichroic mirror reflects the 1083 nm pumping-light back into the cell in order to maximize absorption in the gas. It transmits the 668 nm fluorescence light from the gas discharge whose circular polarization can be measured by a monitor behind the mirror. This polarization is directly correlated to that of the helium. A minimum laser power of about 5 W is required to achieve optimal performance in gas-polarization. Fig. 8 shows the arrangement of the six cells inside the solenoid. As a slightly imperfect circular polarization of the laser beam causes already huge losses in the final ^{3}He polarization [12], birefrigerance has to be avoided. Therefore quartz-windows are glued to both sides of the glass tubes, as quartz does not show the effect of thermal birefrigerance caused by absorption of laser power inside the window.

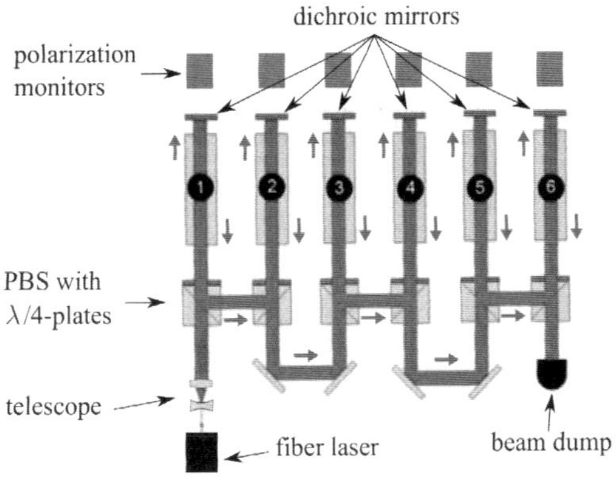

Figure 7. Schematic drawing of the optics and the optical pumping cells. The laser beam passes each glass tube in both directions. The total length at which the beam polarizes gas is around 15 m. Each window is coated in order to reduce reflection power losses to < 1% per surface.

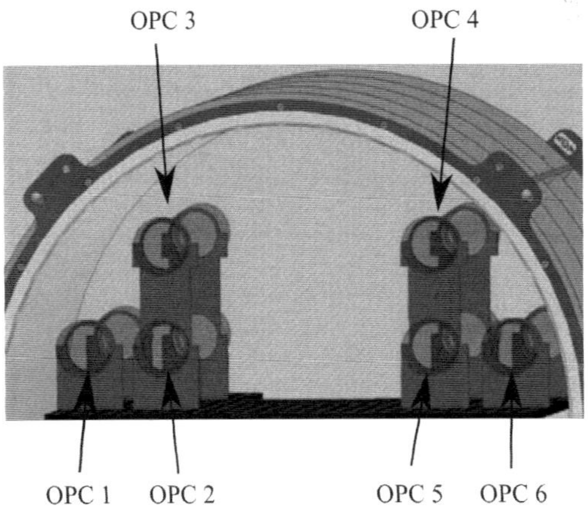

Figure 8. Arrangement of the six optical pumping cells mounted on an optical table inside the solenoid.

Journal of Physics: Conference Series **294** (2011) 012007 doi:10.1088/1742-6596/294/1/012007

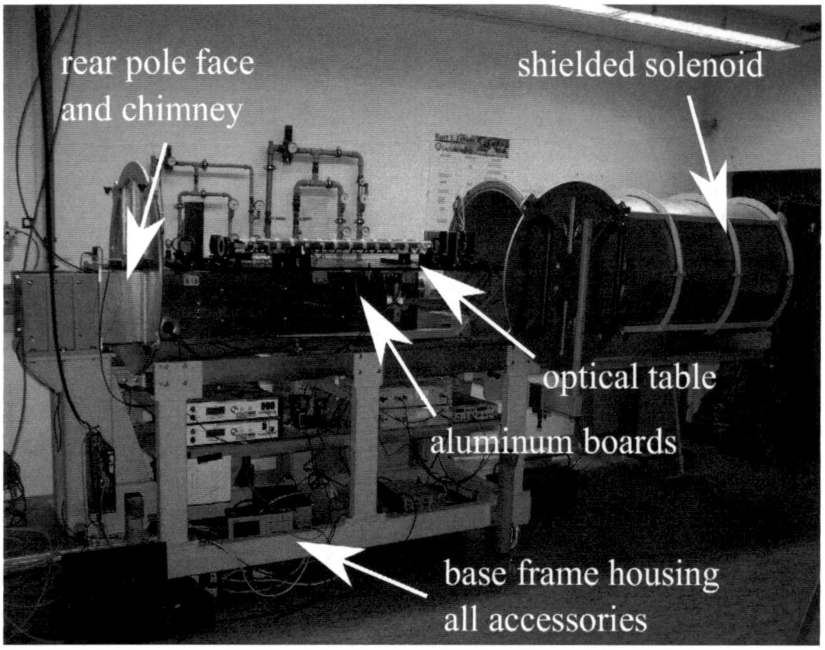

Figure 9. Compact polarizing facilty with shielded solenoid shifted aside.

3.3. The drive for the piston compressor

The compact polarizer is equipped with an electrical linear motor, which drives the piston compressor. Contrary to the current hydraulical drive, this very compact linear drive can be attached directly to the compressor and needs about a factor of two less time for one compression cycle. That higher speed is required, as the compressor volume of the new apparatus is only half of that of the current machine. The linear motor allows a positioning of the piston with the accuracy of one micrometer, which is advanteageous for minimizing the dead volume at the end of the compression phase. The maximum force, that can be generated by the motor is $F_{max} = 11600\,\mathrm{N}$, which allows to compress the gas up to a pressure of $p_{max} = 7.5\,\mathrm{bar}$.

3.4. Mechanical assembly of the new polarizer

The assembly inside the shielded solenoid, in which the gas is polarized is mounted to a pair of aluminum boards which stick out of the chimney at the rear pole face (see fig. 9). They rest on a mike boom whose foot is fixed to the base frame with some freedom for fine adjustment. The solenoid is mounted on telescope guide rails fixed to the base frame. They enable the solenoid to be shifted aside in order to allow easy acess to the assembly inside. The base frame features space for vacuum pumps, electronics, ^{3}He-supply etc. All major construction components outside the shielded solenoid are built from ordinary steel. The compact polarizer measures about 3 m in length, 90 cm in width and 1.8 m in height, which make it fit through a normal door.

3.5. Conclusion

The concepts and results, presented in this article show, that the physical and technical requirements for the construction of a compact polarizing facility are fulfilled. The apparatus, currently being in the final stage of assembling, is designed to provide a high flux of hyperpolarized ^{3}He. The used 15 W fiber laser provides enough power to polarize the gas fast enough to reach a gas throughput of 1 standard liter per hour at P =75%, which is comparable to that of the currently used polarizing facility.

References

[1] Batz M, Baeßler S, Heil W, Otten, E.W, Rudersorf D, Schmiedeskamp J, Sobolev Y and Wolf M 2005 *J. Res. Natl. Inst. Stand. Technol.* B **110** 293
[2] Krimmer J, Heil W, Karpuk S, Otten E.W and Salhi Z 2011 *IOP journal of physics conference series, this volume*
[3] Gemmel C *et al* 2010 *Eur. Phys. J.* D **57** 303
[4] Albert M.S, Cates G.D, Driehuys B, Happer W, Saam B, Springer C.S Jr and Wishnia A 1994 *Nature* **370** 199
[5] Schmiedeskamp J, Heil W, Otten E.W, Kremer R.K., Simon A and Zimmer J 2006 *Eur. Phys. J.* D **38** 427
[6] Hiebel S, Großmann T, Kiselev D, Schmiedeskamp J, Gusev Y, Heil W, Karpuk S, Krimmer J, Otten E.W and Salhi Z 2010 *Journal of Magnetic Resonance* **204** 37
[7] Colegrove F.D, Schearer L.D and Walters G.K 1963 *Phys. Rev.* **132** 2561
[8] Ebert M 2000 doctoral thesis Mainz University
[9] Schmiedeskamp J 2004 doctoral thesis Mainz University
[10] Deninger A, Ebert M, Hasse J, Heil W, Otten E.W, Schmiedeskamp J and Surkau R patent no.s DE 100 00 675 (2000) and [US] 09/758,006 (2001)
[11] The material used is NILOMAG; for specifications see www.specialmetals.com
[12] Wolf M 2004 doctoral thesis Mainz University

Journal of Physics: Conference Series **294** (2011) 012008 doi:10.1088/1742-6596/294/1/012008

An accurate optical technique for measuring the nuclear polarisation of ^{3}He gas

C Talbot[1], M Batz[1,2], P-J Nacher[1*], and G Tastevin[1]

[1] Laboratoire Kastler Brossel, ENS; CNRS; UPMC; 24 rue Lhomond, 75005 Paris, France.
[2] Johannes Gutenberg University, Staudingerweg 7, 55128 Mainz, Germany.

E-mail: nacher@lkb.ens.fr

Abstract. In the metastability exchange optical pumping cells of our on-site production unit and of our other experimental set-ups, we use a light absorption technique to measure the ^{3}He nuclear polarisation. It involves weak probe beams at 1083 nm, that are either perpendicular or parallel to the magnetic field and cell axis, with suitable light polarisations. When metastability exchange collisions control the populations of the sublevels in the 2^3S state, absolute values of the ^{3}He ground state nuclear polarisation are directly inferred from the ratio of the absorption rates measured for these probe beams. Our report focuses on the transverse detection scheme for which this ratio, measured at low magnetic field for σ and π light polarisations, hardly depends on gas pressure or the presence of an intense pump beam. This technique has been systematically tested both in pure ^{3}He and isotopic mixtures and it is routinely used for accurate control of the optical pumping efficiency as well as for calibration of the NMR system.

1. Introduction

Metastability exchange optical pumping (MEOP) allows the polarisation of ^{3}He nuclear spins in pure helium gas and this technique is successfully used for production of highly polarised ^{3}He with large throughput. MEOP requires the electronic excitation of a small fraction of the He atoms to the metastable (2^3S) state and involves two co-operative processes: strong selective excitation on the 2^3S-2^3P optical transition line at 1083 nm with transfer of angular momentum from polarised light to metastable He atoms; metastability exchange collisions with, for ^{3}He, transfer of nuclear polarisation from the metastable state to the ground state [1]. In pure ^{3}He gas, the driven radiative transitions enforce simultaneous orientation of the electronic and nuclear spins in the 2^3S state, entangled by the hyperfine interaction. In isotopic gas mixtures, radiative excitation can selectively enforce orientation of ^{4}He electronic spin in the metastable state but 2^3S polarisation builds up also for the ^{3}He isotope through metastability exchange between ground state ^{3}He and optically pumped ^{4}He atoms. For a detailed description of the MEOP process, we refer the reader to the companion contribution on MEOP basics in the same conference proceedings [1] as well as to comprehensive regular articles [2, 3, 4].

Radiofrequency (rf) discharges are frequently used for electronic excitation of a small (ppm) fraction of He atoms to the metastable state and MEOP performs better at low or moderate pressures (typically 0.5$-$5 mbar for operation at mT field strengths and 5$-$250 mbar for operation between 0.4 T and 4.7 T). Gas production units thus typically include a combination of several systems for high-purity gas handling, high-power optical pumping, polarisation preserving compression, and long-term or short-term storage for massive centralised production or flexible on-site production, respectively. Diagnostic systems are also implemented for monitoring of the operating conditions and optimisation of the performances of the gas production units. NMR or magnetometry can be used to measure polarisation in the storage cell where large magnetisation is accumulated. In the optical pumping cell, optical methods such as polarimetry of the visible fluorescence light emitted by the plasma [5, 6, 7] or 1083 nm light absorption measurements [2, 4, 8, 9] are preferred because they provide much higher sensitivity and operate in the presence of the rf discharge.

All these techniques require calibration for accurate polarisation measurements, except the light absorption technique. The latter has actually been used for calibration purposes in the past decades [10, 11]. In contrast with the visible light polarimetry technique, that becomes inefficient above approximately 10 mT due to strong hyperfine decoupling in the highly excited energy levels involved in the radiative cascade, the absorption technique potentially operates at all field strengths [4]. It has been successfully used for high field MEOP investigations [4, 12, 13]. At low magnetic fields, recent work [14] has motivated more quantitative investigations of its potential artefacts in the context of very high power optical pumping.

In this contribution, we describe an accurate way to measure the ^{3}He nuclear polarisation at low magnetic field with the absorption technique, in various MEOP conditions. We report on systematic dedicated tests performed in small sealed He cells filled with pure ^{3}He gas or isotopic gas mixtures, we address the practical aspects of the experimental implementation, and quantitatively discuss the most relevant identified sources of errors in polarisation measurements.

2. Polarisation measurements with the light absorption technique

The absorption technique is based on the very strong link between the ground state nuclear polarisation M and the distribution of populations between the three (for ^{4}He) and six (for ^{3}He) Zeeman sublevels in the metastable state that is established by the frequent exchange collisions between the 2^3S atoms and the much more numerous ground state atoms. When metastability exchange dominates, a spin temperature distribution is enforced and the populations of the 2^3S sublevels scale exponentially with their respective angular momentum with a Boltzmann parameter $e^{\beta}=(1+M)/(1-M)$ [4]. The polarisation M can then be simply inferred from the ratio of populations in two 2^3S sublevels with distinct angular momenta for either isotope, or from independent linear combinations of such populations.

At low magnetic field, the 2^3S Zeeman sublevels are almost degenerate and two absorption measurements can be simultaneously performed with a monochromatic light source using different light polarisations. Three 2^3S-2^3P$_0$ lines are well resolved in a low-pressure gas: the C$_8$ line (2^3S, F=1/2-2^3P$_0$) and the C$_9$ line (2^3S, F=3/2-2^3P$_0$) in pure ^{3}He, and the ^{4}He D$_0$ line (2^3S-2^3P$_0$) in isotopic mixtures. Picking up one of these lines, one may choose among the following simple schemes:

1/ Transverse probe: a weak laser beam propagates perpendicular to the axis of the magnetic field B and transmitted powers are measured for the orthogonal linear polarisations π (parallel to B) and σ (perpendicular to B). M is inferred from the ratio $r_{\perp}=A_{\pi}/A_{\sigma}$ of the corresponding absorption signals (see Section 3.5) A_{π} and A_{σ}, respectively. At zero magnetic field, the ground state ^{3}He nuclear polarisation is given by:

$$M=\pm\sqrt{\frac{1-r_{\perp}^{9}}{1+2r_{\perp}^{9}}}\quad \text{for C}_9, \tag{1}$$

$$M=\pm\sqrt{\frac{1-r_{\perp}^{0}}{1+r_{\perp}^{0}}}\quad \text{for D}_0. \tag{2}$$

These formulas provide good approximations at small B if the probe is tuned to the centre of the absorption Doppler profile. For example, errors in M are lower than 2.5×10^{-3} for $|M|>0.1$

and B=1 mT. For higher magnetic fields, they can still be used provided that $r_\perp$ is replaced by the normalised ratio $R_\perp = r_\perp(M)/r_\perp(0)$ obtained by scaling $r_\perp$ to its value at null polarisation, $r_\perp(0)$. If the probe laser is not tuned to the centre of the null-field or null-polarisation absorption profile, the detuning introduces a field-dependent error on $|M|$ of order -0.02/GHz/mT.

Polarisation cannot be measured with probe light tuned to C_8 because this line addresses the same pair of 2^3S sublevels (F=1/2, m_F=±1/2) for π and σ polarisations, hence the ratio $r_\perp^8$ does not depend on M. The C_9 line cannot be used for polarisation measurements in mixture cells because it overlaps with the strong unresolved D_1 and D_2 lines ($2^3S{-}2^3P_1$ and $2^3S{-}2^3P_2$, respectively) of the ^{4}He isotope. Therefore, absorption measurements are performed with the probe laser tuned to the C_9 line in pure ^{3}He gas and to the D_0 line in isotopic mixtures.

2/ Longitudinal probe: the weak laser beam propagates parallel to B, transmitted powers are measured for the two circular polarisations σ_+ and σ_-, and M is inferred from the ratio $r_\parallel = A_{\sigma+}/A_{\sigma-}$ of the absorption signals $A_{\sigma+}$ and $A_{\sigma-}$, respectively. At zero magnetic field, the ground state ^{3}He nuclear polarisation is given by:

$$M = \frac{1 - r_\parallel^8}{1 + r_\parallel^8} \quad \text{for } C_8, \tag{3}$$

$$M = \frac{1 - \sqrt{r_\parallel^0}}{1 + \sqrt{r_\parallel^0}} \quad \text{for } D_0. \tag{4}$$

The polarisation components σ_+ and σ_- individually address a single sublevel for these two single-component lines (m_F=−1/2 and +1/2, respectively, for C_8 and m_J=−1 and +1, respectively, for D_0) and the ratio $r_\parallel$ directly measures at B=0 the ratio of populations in the probed Zeeman sublevels. As is done for the transverse probe, $r_\parallel$ can be replaced by the normalised ratio $R_\parallel = r_\parallel(M)/r_\parallel(0)$ in formulas (3) and (4) to obtain accurate results in an applied field. Moreover, the use of reduced ratios also eliminates the error introduced by probe detuning for the longitudinal probe scheme.

No simple analytical formula exists for the C_9 line for which each polarisation component addresses simultaneously two Zeeman sublevels. Instead, the relation:

$$r_\parallel^9 = \left(\frac{2-M}{2+M}\right)\left(\frac{1-M}{1+M}\right)^2 \tag{5}$$

has to be solved (e.g., finding the root of a cubic polynomial, numerically, or using polynomial fits...) to infer M from $r_\parallel^9$.

Figure 1 shows the variations of $r_\perp$ and $r_\parallel$ with M. For the transverse probe scheme, C_9 and D_0 lines provide a comparable sensitivity for polarisation measurements, quite high above $|M| \geq 0.2$. The sensitivity is reduced at small M, with a quadratic departure of $r_\perp$ from 1 that results from the fact that each light component simultaneously addresses sublevels with opposite angular momenta. In contrast, the longitudinal probe scheme has a linear response at small M for all lines, with a higher sensitivity for the C_9 and D_0 lines. At high M, it has a reduced sensitivity for C_9 and D_0 with $r_\parallel \propto (1-M)^2$; the C_8 line retains a linear sensitivity, yet provides a limited precision since both absorption signals vanish at M=1.

During optical pumping, a strong pump beam tends to enforce a distribution of populations in the 2^3S states that departs from the spin-temperature distribution that would result from metastability exchange only. The above equations then provide apparent polarisations M^{app} that depend on the choice of detection scheme and probe line and differ from the actual nuclear polarisation M. Examples of experimental deviations of M^{app} from M are discussed in Section 5.

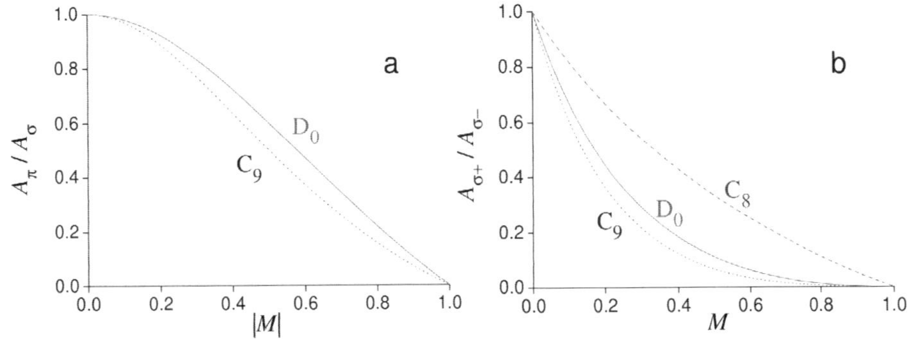

Figure 1. Ratios of 2^3S-2^3P$_0$ absorption signals computed for the transverse (a) and longitudinal (b) probe schemes as a function of ground state nuclear polarisation M, at zero magnetic field and at spin temperature equilibrium. **a-** Transverse probe scheme:. D_0 (solid line) and C_9 (dotted line) provide comparable sensitivities at all M. Opposite polarisations M and $-M$ cannot be distinguished using A_π/A_σ. **b-** Longitudinal probe scheme: C_9 (dotted line) and D_0 (solid line) provide higher sensitivity than C_8 (dashed line) at small M, where accurate measurements can be performed in contrast with the transverse scheme. Identical curves would be obtained by plotting the inverted ratio $A_{\sigma-}/A_{\sigma+}$ for negative M values.

3. Experimental apparatus

3.1. Probe laser

The probe laser consists of a linearly polarised 70 mW DFB laser diode (Toptica Photonics AG) housed within a temperature-regulated aluminium box for wavelength stability. Inside this box the divergent elliptical beam from the laser diode is collimated and then reshaped to circular with a pair of anamorphic prisms. The beam passes through a Faraday isolator before being coupled into a polarisation preserving monomode optical fibre as it exits the aluminium box. The laser can thus be situated remotely to the helium polarising equipment for both convenience and reduced impact on the magnetic field map. Typically 15 milliwatts are coupled into the fibre when the laser diode is running at maximum current. Laser stability and control are achieved with an in-house electronic controller. The aluminium box is heated to a fixed set point above room temperature with a resistive element for long term thermal stability and the laser frequency is adjusted using the on-chip thermoelectric cooler.

3.2. Probe assembly for the transverse probe scheme

Figure 2 shows a schematic diagram of the probe assembly and its caption includes all technical details. The laser diode light that exits the optical fibre tip is collimated and split in two adjacent beams with σ and π polarisations. This is in contrast with prior implementation [10], in which probe beams spatially overlapped. The beam separation preserves the accuracy of the polarisation measurements for all probe tunings, as discussed in Section 5.1. The beams propagate across the helium cell, entering and exiting through the curved glass walls. A cylindrical lens is used to correct for the resulting wavefront distortions before back-reflection through the lens and cell. One of the beams is then deflected by a polarising beam splitter (PBS) cube. Both beams are coupled by lenses into 1-mm core multimode plastic fibres that are connected to the photodiodes. The photodiode voltages are recorded on a PC using a USB data logger for post processing.

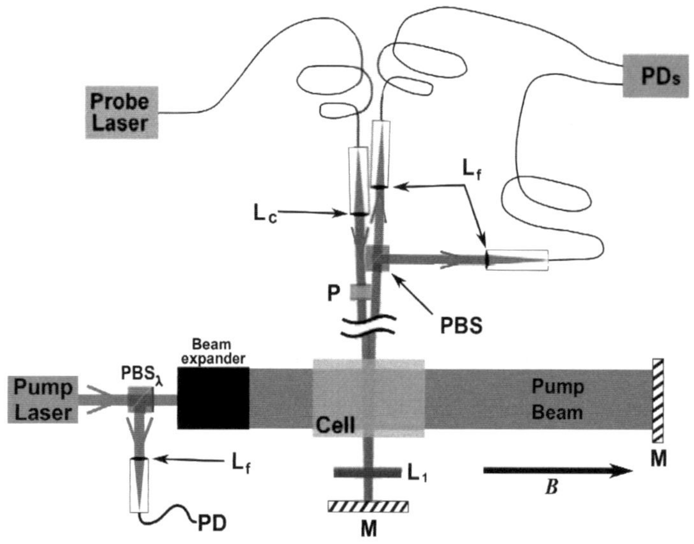

Figure 2. Schematic of the transverse probe assembly. Optical elements include: mirrors (M), photodiodes (PD); a collimating lens (L_c, f=40mm), a polarising cube (PBS), and a compensating cylindrical lens (L_1, f=500mm), and collecting lenses (L_f, f=30mm) on the probe path; a circular polariser (PBS_λ, a PBS cube with glued quarter wave plate) and a collecting lens (L_f, f=30mm) on the pump path. The collimated laser diode probe beam is split in two halves by P, a pair of co-planar polarising plates with crossed linear axes attached to a mask that consists in a 8 mm circular hole with a central 1.5-mm broad beam stopper. The exiting collinear beams with polarisations π (along B) and σ (perpendicular to B) cross the cell at almost normal incidence (2° tilt). The distance between cell axis and probe beam collection is approximately 50 cm.

3.3. Pump laser

For optical pumping a 5 Watt broadband (1.7 GHz fwhm) 1083 nm fibre laser is used (Keopsys, model YFL-1083-50-COL). The collimated Gaussian output beam is circularly polarised and expanded to a diameter of 2 cm fwhm. It propagates along the direction of the magnetic field and passes twice through the cell. The transmitted light, deflected by the circular polariser, is monitored by a photodiode for checks of the laser tuning and measurements of the pump power absorption.

3.4. Helium test cells

The sealed helium cells used for these investigations are pyrex glass tubes with flat end windows (5 cm in diameter 5 cm long). They contain ^{3}He at various gas pressures, either pure or mixed with set ratios of ^{4}He. However some of the "pure ^{3}He" cells actually contain traces of ^{4}He plausibly left from the cleaning process prior to the filling of the cells.[1] A discharge is generated in the cell using two external wire electrodes connected to a tuned voltage transformer fed by a rf generator able to produce 5 W in the 1 to 3 MHz range. This rf excitation is amplitude modulated at around 70 Hz and lock-in detection is used to improve sensitivity of the absorption measurements.

[1] ^{4}He concentrations, measured by comparison of absorption rates on the resolved C_8 and D_0 isotopic lines, range from 0.1 to 2 %.

3.5. Absorption signals

The absorption signals A_π and A_σ used for nuclear polarisation measurement are the amplitudes of the modulation depth (i.e., the ratios of the ac amplitudes to the dc components) of the corresponding probe powers exiting the cell. Each ratio is robust against fluctuations of the incident power I_0 and directly proportional to the absorbance $-ln\,(I/I_0)$, with a coefficient that only depends on the plasma response to the rf excitation [4, 15]. For improved accuracy, the PBS cube (transmission efficiency: 1-ϵ=0.96) is oriented so as to transmit the π component for which the absorbance is very small at high nuclear polarisation ($A_\pi \longrightarrow 0$ for $M \longrightarrow 1$) and to deflect the σ component. In this configuration, the amplitudes S_π and S_σ of the modulation depth of the photodiode voltages are related to the absorption signals by S_π=(1-ϵ)A_π and S_σ=A_σ+ϵA_π. This is used to compute the ratio $r_\perp$ of absorption signals from processed photodiode voltage data:

$$r_\perp = A_\pi/A_\sigma = \frac{S_\pi}{S_\sigma - \epsilon S_\pi/(1-\epsilon)}. \tag{6}$$

3.6. NMR detection

For comparison with optical measurement results, NMR measurements have been performed at 28 kHz (the Larmor precession frequency for B=0.87 mT) using a homebuilt NMR detection system that consists of crossed induction and detection coils set around the test cell. Since the gas density is low, 90° tipping pulses are routinely used to obtain good signal-to-noise ratios (typically up to 100), with tens of seconds of precession lifetimes T_2. NMR signals are proportional to the fixed number N_3 of ^{3}He atoms contained in the glass cell and to the nuclear polarisation M. They have to be scaled for cell content and volume for direct comparison to the nuclear polarisations measured using the absorption technique.

The gas pressures slightly differ from the filling pressures for the sealed cells that are several decades old and the actual pressures are measured by NMR. For each cell, the decay rates of free precession signals, $1/T_2$, have been measured for various applied magnetic field gradients to obtain the ^{3}He diffusion coefficient D [16]. D is inversely proportional to gas pressure P for pure ^{3}He gas (D=1.997$\times[T/300]^{1.71}/P$, with temperature T in K and P in atm) and depends on the partial pressures of ^{3}He and ^{4}He gas for isotopic mixtures (see reference [17], and corrected values of diffusion coefficients in [18]).

4. Polarisation measurements

Figure 3 shows probe recordings and processed data from a typical polarisation experiment. During the acquisition, the photodiode voltages and TTL reference signal are continuously recorded. Figure 3a shows a small section of the recorded transmitted probe powers in which the modulation of the absorption is clearly seen. Initially (part C) the gas is not polarised (M=0) and the modulation depth of the π and σ components are equal; later (part D) the gas is polarised (M=0.5) and both absorption rates and modulation depths are different for the two components. Figure 3b shows the time evolution of the average absorption signals and computed nuclear polarisation during the entire experiment. The plot is divided into six regions that are related to the six main steps of the acquisition protocol, as follows:

(A) the laser beams are blocked and the rf discharge is off; offset voltages are recorded.

(B) the probe block is removed and discharge is off; incident probe powers I_0 are measured.

(C) the rf discharge is on; transmitted probe powers I are measured at null polarisation.

(D) the pump block is removed; optical pumping takes place and polarisation increases.

(E) the pump beam is blocked again when polarisation is close to steady-state; polarisation decreases due to plasma-induced decay for some arbitrary time period.

(F) the rf discharge is turned off; NMR signal is acquired to measure the magnetisation associated with the remaining polarisation M.

Nuclear polarisation is fully destroyed in the cell by the 90° rf pulse used for the NMR measurement. MEOP is performed to repolarise the gas and another measurement can be made for a different polarisation M after a shorter or longer decay period.

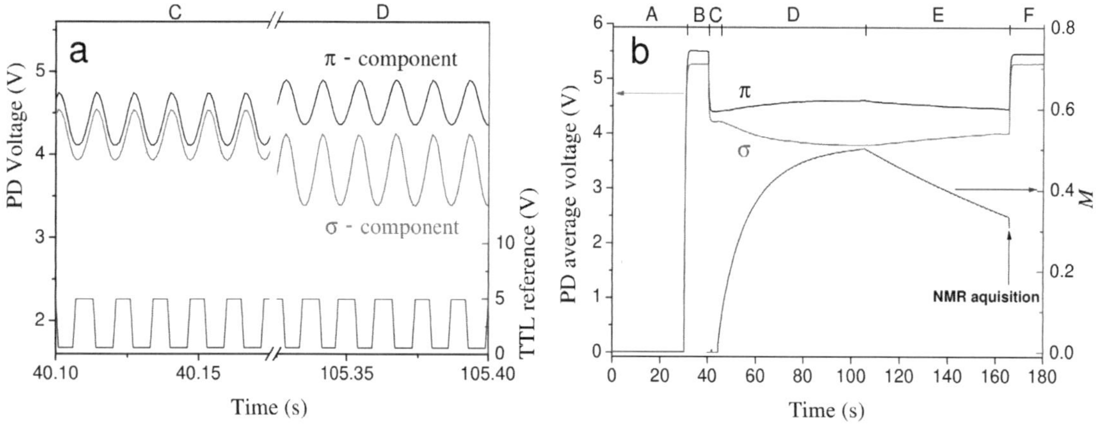

Figure 3. Example of raw and processed data from a typical polarisation experiment performed on the 1.3 mbar cell. The probe laser is tuned to the C_9 transition with 0.14 mW of output power. The polarisation is achieved by MEOP using the pump laser tuned to C_8 with 2.5 W of output power. Fig. 3a shows two subsections of the raw voltage recordings (upper traces: PD voltages, lower trace: TTL reference signal). Section C: discharge is on, pump is blocked, and transmitted probe powers are recorded at M=0, showing equal modulation depths (~10%). End of section D: after 60 s of MEOP, the ratio of the modulated absorptions is significantly modified. Fig 3b shows the time averaged probe signals (upper traces) and the computed M (lower trace) for an entire polarisation experiment. The signal-to-noise ratio exceeds 3000 at M=0.5 for an averaging time constant of 0.3 s.

Excellent linear correlation is obtained between the two independent sets of polarisation measurements (NMR and optical probe). In all the pure ^{3}He cells the absorption technique is checked to yield data that consistently agree with the NMR data scaled to actual gas pressures. In the mixture cells, the absorption technique using the D_0 line yields data that also fully agree with the NMR data scaled to ^{3}He partial gas pressures. This agreement confirms that the nuclear polarisation of ground state ^{3}He atoms imposes a spin temperature distribution in the ^{4}He metastable state as well.

5. Discussion

5.1. Implementation and design

The absorption measurements are performed on the same infrared transition as the optical pumping. Contribution to the measured transmitted powers from the 1083 nm fluorescence light emitted by the gas subjected to intense pumping should thus be limited, as usual, by reducing the detected solid angle to a minimum. Alternatively, when fluorescence light and stray light from the intense pump laser cannot be avoided, a double modulation scheme (e.g., with a probe frequency swing in and out resonance or mechanical beam chopping) may help. The stray light contribution is a minor issue for the transverse probe scheme, when compared to the longitudinal one. However the absorbed probe power (hence the signal-to-noise ratio) is generally lower with transverse probe scheme, especially in the long cells used for gas production.

Journal of Physics: Conference Series **294** (2011) 012008 doi:10.1088/1742-6596/294/1/012008

In order to measure absolute ratios of populations, the two probe components should ideally overlap exactly. However, in this situation, polarisation components reaching the same 2^3P Zeeman sublevel may interact. The scaled ratio of absorption signals $R_\perp = r_\perp(M)/r_\perp(0)$ has been observed to exhibit small frequency-dependent variations at finite nuclear polarisations, both when the probe is tuned to the C_9 line in ^{3}He gas and when it is tuned to the D_0 line in isotopic gas mixtures. Similar observations have been made with the longitudinal probe scheme for $R_\| = r_\|(M)/r_\|(0)$ in pure ^{3}He gas when absorption measurements are performed on the C_9 line, but no such artefact has been noticed when the C_8 line is used. Systematic investigations have been made for both probe schemes in a variety of experimental conditions and detailed results will be reported elsewhere. Using two separate beams for the absorption measurements, as described in Section 3.2 for the transverse probe scheme, has been checked to make the optical technique robust against probe detuning and accurate at all polarisations (for both probe schemes). These probe beams still remain very close so as to address nearly identical sets of atoms, and any small local differences (e.g., in the number density of metastable atoms) cancel out when reduced ratios $R_\perp$ or $R_\|$ are used.

5.2. Impact of the probe light intensity

The light intensity in the probe beam must be low enough so as not to induce significant distortion of the 2^3S population distribution in the metastable state [4]. This distortion may be expected to scale as γ_p/γ_e in pure ^{3}He gas, where γ_e is the metastability exchange rate and γ_p is the relevant average optical pumping rate for the probed line components. γ_e is proportional to gas pressure P (γ_e=3.75×10^6 s^{-1}/mbar for ^{3}He [19]) and γ_p is proportional to the light intensity and scales as $1/P$, due to the impact of velocity-changing collisions on the fraction of the atomic Doppler profile actually addressed by the single-frequency probe light [3]. We have performed checks of this effect by measuring the ratios $r_\perp^9$ and $r_\perp^0$ for M=0 in two ^{3}He cells and in a mixture cell, respectively. They decreased linearly with decreasing probe intensity, with slopes consistent with expectations and obeying the $1/P^2$ scaling for pure ^{3}He.

The light intensity should be reduced until it does not affect the measurements. For instance, for a probe intensity of 100 μW/cm^2, M is overestimated at low pressure (P=0.43 mbar) by as much as 0.05 for M=0 and 0.01 for M=0.1. These errors are reduced to 0.023 (M=0) and 0.002 (M=0.1) for a probe intensity of 25 μW/cm^2. It is possible to keep the intensity of the probe low by reducing the incident power and/or increasing the beam diameter. The second option is usually preferred because it preserves the signal-to-noise ratio. These limits on probe intensity are significantly less stringent if a broadband laser is used, such as the monitor output of the fibre laser used for optical pumping. The actual distortion of population distribution is then reduced by the ratio of linewidth (of order 100) at given laser intensity, hence intensities of order 1 mW/cm^2 may be used.

5.3. Impact of the probe light power

The probe-driven optical pumping cycles may also influence the time evolution of nuclear polarisation because they contribute to the overall angular momentum budget [1]. No angular momentum is associated to the π and σ components of the transverse probe but, since the probe light equally addresses sublevels with opposite angular momenta, it tends to more efficiently depopulate the overpopulated ones. As a result it drives the 2^3S state towards a null average nuclear orientation and, through metastability exchange, contributes to the ^{3}He polarisation decay in the ground state. This may limit the MEOP performances in optical pumping experiments, in a way that is similar to what occurs for an imperfectly polarised pump beam. This may also influence the evolution of the polarisation in the absence of pump light and introduce a noticeable additional decay rate. The additional decay rate is expected to scale with the absorbed power (hence, with the 2^3S number density) and to be inversely proportional to the number of ground state atoms (hence, to the ^{3}He gas partial pressure and to the cell volume). This has been experimental checked by recording polarisation decays in the plasma for various probe beam powers. Figure 4 shows that decay rates increase linearly

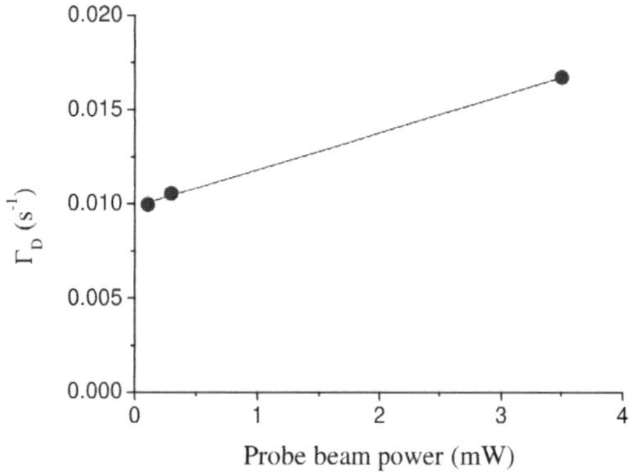

Figure 4. Examples of polarisation decay rates plotted as a function of probe power in the transverse probe scheme. The slope indicates a probe-induced additional decay rate of 2×10^{-3} s^{-1}/mW, consistent with expectations for these experiments (0.43 mbar cell, 9% probe absorption).

with probe powers at fixed rf discharge excitation, as expected. Depending on the objective of the polarisation measurement, a trade-off between a high signal-to-noise ratio and a negligibly small action of the probe has to be found; probe powers in the range 0.1-1 mW are usually a safe choice. When long cells are used, the increase in their volume proportionally decreases the impact of probe power, but in the transverse scheme only.

5.4. Impact of the pump light

The deviation from spin temperature distribution of populations induced by the strong pump light, that provides the driving term for the build up of the ground state ^{3}He polarisation, results in modified absorption rates for the two probe beams [1]. The apparent polarisation values M^{app} obtained using the relevant spin temperature formula (Eqs. (1) to (5), depending on the detection scheme) overestimate the actual polarisations M and the pump-induced differences M^{app}-M can be experimentally measured when the pump is rapidly applied or blocked, at steady-state or at arbitrary M during polarisation build-up. Results of dedicated experiments performed to quantify this difference for both probe schemes in similar OP conditions are shown in Figs. 5 and 6.

Data points in Fig. 5 have been individually obtained using the protocol described in Section 4 for various polarisation growth times. The uncertainties mainly result from polarisation changes during the time required to manually block the OP beam. The pump-induced differences decrease when gas pressure rises (from 0.43 to 5.4 mbar), due to the increased metastability exchange rate that competes with optical pumping for population distribution, and steadily decrease with high nuclear polarisations at all pressures. When the same line is used for OP and for detection, the pump and probe beams address the same 2^3S Zeeman sublevels and the pump-induced differences are observed to be about twice as large as in Fig. 5 (data not shown).

Each of the four data sets in Fig. 6 has been obtained from a single recording, where short periods of polarisation build-up alternated with periods with a blocked pump. For each choice of probe line, the pump-induced differences systematically decrease again with pressure and polarisation, and they increase by a factor of five when the C_8 line is used instead of the C_9 line for the probe. When the C_9 line is used for pumping (data not shown), the pump-induced differences for each pressure are similar for the two probe lines and about twice as large as those represented by the solid symbols in Fig. 6.

Comparison of Fig. 5 to Fig. 6 shows that the transverse probe scheme is significantly less prone to the impact of pump light than the longitudinal probe scheme (5 to 10 times less,

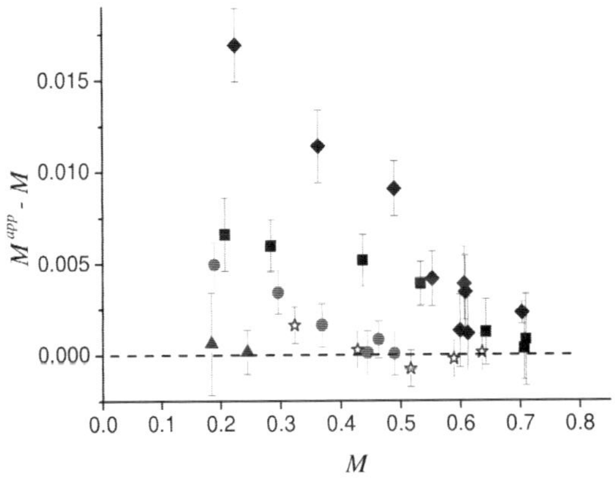

Figure 5. Measured pump-induced differences M^{app}-M for the transverse probe scheme at various pressures and polarisations. OP is made with 1.5 W laser output power tuned to the C_8 line. Solid symbols: C_9 probe, pure ^{3}He gas, (diamonds: 0.43 mbar; squares: 1 mbar; circles: 1.33 mbar; triangles: 5.41 mbar). Open stars: D_0 probe, mixture cell (1 mbar ^{3}He and 0.47 mbar ^{4}He).

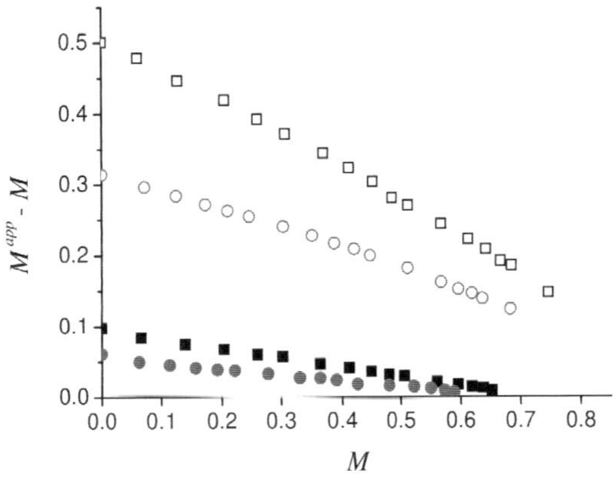

Figure 6. Measured pump-induced differences for the longitudinal probe scheme (note the difference in vertical scales with Fig. 5). OP is made with 2 W laser output power tuned to the C_8 line (beam diameter: 1.4 cm fwhm) in 6 cm diameter 30 cm long cells filled with pure ^{3}He gas (squares: 0.63 mbar; circles: 1.19 mbar). Solid symbols: C_9 probe; open symbols: C_8 probe. The schematic of the longitudinal probe assembly used for the measurements is given in [1].

depending on the chosen OP transition). This is due to partial compensation of the effects on the transverse absorption signals of the pump-driven depopulation and overpopulation for pairs of 2^3S sublevels with opposite angular momenta. For both detection schemes, pump-induced differences increase with laser power until optical saturation of the 2^3S-2^3P$_0$ occurs (typically above 1W/cm^2).

In previous work using a transverse probe scheme [10], accurate comparisons of the absorption and polarimetry techniques had been performed without pump light, during polarisation decay. The reported absence of noticeable difference between the polarisations measured with (M^{app}) and without (M) pump beam explicitly refers to comparisons performed at steady-state polarisations. It can be explained by the limited signal-to-noise ratio (Fig. 4 of [10]) obtained with the single-pass probe path and by the high nuclear polarisations achieved with a broadband 3W laser ($M \geq 0.5$).

For standard operating conditions, the nuclear polarisation exceeds 0.5 in the optical cells of our gas production unit and the steady-state pressure is on the order of 1.33 mbar. Overlooking the small difference between M^{app} and M results in an absolute error smaller than 0.25% and relative errors systematically lie below 0.5% when the transverse probe scheme is used.

5.5. Impact of ^{4}He impurities

Residual ^{4}He atoms may contribute to probe light absorption due to the overlap of the D_1 and D_2 lines with the C_9 line ^{3}He atoms. Computations performed for the upper bound of ^{4}He concentrations measured in the sealed test cells (2%) yield, for instance, M^{app}=56% for M=60%. The relative error approximately scales with the ^{4}He concentration .

6. Conclusions

We have reported implementation and systematic tests of the transverse probe scheme for absolute polarisation measurements with the absorption technique. The consistency and high accuracy of the measurements have been demonstrated and assessed by correlation with NMR measurements as well as by comparison with numerical results obtained with a MEOP model. Potential systematic errors have been discussed and a number of simple experimental tests have been described to check that they actually remain well below the percent level.

Most experiments have been performed using a single-frequency laser diode to probe 2^3S-2^3P_0 absorption. The technique can be applied also using a broadband light source. The overall decrease in probe light absorption has no influence on the computed M values, but data reduction becomes less straightforward as some contribution to light absorption comes from the wings of neighbouring resonance lines for very broad lasers. We have performed test measurements with the weak monitor beam provided by the supplementary output of the pump laser, but low accuracy has been obtained with this more noisy light source. Moreover, its use is restricted to situations where MEOP is performed using the C_9 or D_0 line.

The probe assembly that we have described has also been used for polarisation measurements in the long valved cell (50 cm length, 6 cm diameter) of our gas production unit to characterise the efficiency of MEOP in various operating conditions. For improved performance, a double cell with two such glass cylinders connected in series has been recently implemented in the unit. The same transverse probe assembly is currently used for simultaneous polarisation measurements in the two gas compartments. Successful measurements are performed with a configuration similar to that of Fig. 2 where the probe beam propagates successively through the two glass cells before back-reflection by the mirror. Using two different frequencies for the discharge modulation in each cell, their individual contributions to the absorption signal can be singled out by selective lock-in detection. In spite of the lower signal-to-noise ratios obtained with the transverse probe scheme, the simpler technical implementation makes this scheme quite advantageous over the longitudinal one for polarisation measurements in multiple cells with a single laser beam.

The transverse probe scheme provides a very robust way to perform absolute measurements of the ^{3}He nuclear polarisation in all situations where the gas pressure is not well known or variable in time. Gas production units operate either in batch mode (no gas flow) or in continuous mode (gas flow through the optical pumping cell where fresh gas gets polarised), in general with no easy and non-relaxing way to measure on-line the pressure inside the MEOP cell. The longitudinal probe scheme is more sensitive at low polarisations, but dedicated experiments must be performed to establish the pressure-dependent correspondence between M^{app} and M. The use of this scheme may be restricted to batch mode operation or to reproducible stationary conditions of gas flow. The transverse probe scheme is therefore the best suited one for accurate on-line polarisation measurements.

Acknowledgments

The authors wish to thank Raphaël Lambert for his contribution to an early stage of the work. This research program is supported in part by the European Commission (PHeLINet Marie Curie Research and Training Network grant).

References

[1] Batz M, Nacher P J and Tastevin G 2011 *IOP journal of physics conference series, this volume*
[2] Colegrove F D, Schearer L D and Walters G K 1963 *Phys. Rev.* **132** 2561–72
[3] Nacher P and Leduc M 1985 *Journal de Physique* **46** 2057–73
[4] Courtade E, Marion F, Nacher P J, Tastevin G, Kiersnowski K and Dohnalik T 2002 *European Physical Journal D* **21** 25–55
[5] Laloë F 1971 *Annales de Physique* **6** 5–69
[6] Pavlovic M and Laloë F 1970 *Journal de Physique* **31** 173–94

[7] Stoltz E, Villard B, Meyerhoff M and Nacher P 1996 *Appl. Phys. B: Lasers and Optics* **63** 635–40
[8] Greenhow R C 1964 *Phys. Rev.* **136** A660–2
[9] Daniels J and Timsit R 1971 *Canadian Journal of Physics* **49** 539
[10] Bigelow N P, Nacher P J and Leduc M 1992 *Journal de Physique II* **2** 2159–79
[11] Lorenzon W, Gentile T R, Gao H and McKeown R D 1993 *Phys. Rev. A* **47** 468–79
[12] Abboud M, Sinatra A, Maitre X, Tastevin G and Nacher P J 2004 *Europhysics Lett.* **68** 480–6
[13] Suchanek K, Suchanek M, Nikiel A, Palasz T, Abboud M, Sinatra A, Nacher P J, Tastevin G, Olejniczak Z and Dohnalik T 2007 *European Physical Journal - Special Topics* **144** 67–74
[14] Batz M 2011 *Metastability exchange optical pumping in ^{3}He gas up to 30 mT: Efficiency measurements and evidence of laser-induced nuclear relaxation* Ph.D. thesis UPMC - Paris 6 and Johannes Gutenberg-Universität Mainz to be published on URL `http://tel.archives-ouvertes.fr/`
[15] Courtade E 2001 *Pompagc optique de l'hélium dans des conditions non-standard (See Appendix 1)* Ph.D. thesis Université Pierre et Marie Curie - Paris 6 URL `http://tel.archives-ouvertes.fr/tel-00001447`
[16] Hayden M E, Archibald G, Gilbert K M and Lei C 2004 *J. Magn. Resonance* **169** 313–22
[17] Bidinosti C P, Choukeife J, Nacher P J and Tastevin G 2003 *J. Magn. Reson.* **162** 122–32
[18] Tastevin G and Nacher P J 2005 *J. Chem. Phys.* **123** 064506
[19] Dupont-Roc J, Leduc M and Laloë F 1971 *Phys. Rev. Lett.* **27** 467–70

Journal of Physics: Conference Series **294** (2011) 012009 doi:10.1088/1742-6596/294/1/012009

An active feedback scheme for low field NMR experiments

E Baudin[1], K Safiullin[1], S W Morgan[1,2] and P-J Nacher[1*]

[1] Laboratoire Kastler Brossel, ENS; CNRS; UPMC; 24 rue Lhomond, 75005 Paris, France.
[2] Laboratoire Structure et Dynamique par Résonance Magnétique, IRAMIS, CEA, CNRS, UMR 3299 SIS2M, CEA/Saclay, 91191 Gif-sur-Yvette, France.

E-mail: nacher@lkb.ens.fr

Abstract. In low field nuclear magnetic resonance (NMR) it is desirable to combine proper characteristics of the detection scheme with a good signal to noise ratio. For example, a reduced coupling between the sample and the detection coil is needed for NMR with highly magnetized samples and a large bandwidth is required in magnetic resonance imaging (MRI). We discuss a solution based on a simple active feedback circuit that preserves the signal to noise ratio as opposed to traditional solutions which do not. We give illustrations of its use in experiments on low temperature hyperpolarized liquid ^{3}He-^{4}He mixtures and in hyperpolarized ^{3}He gas MRI.

1. Introduction

Nuclear magnetic resonance (NMR) detection schemes are usually based on resonators tuned to the Larmor precession frequency f_L which are strongly coupled to the sample (with a large geometrical filling factor). A high quality factor (Q-factor Q) of the resonator provides a Q-fold signal enhancement and contributes to a high signal-to-noise ratio (SNR) of the recordings.

Though achieving a high SNR is desirable, the high Q-factor has the drawback of enhancing the interaction between the sample and the detection coil, leading to (often) unwanted radiation damping: the emf induced by the precessing magnetization drives a rf current in the resonator at precisely the Larmor frequency, and the oscillating magnetic field associated with this current tends to drive the magnetization into its stable orientation along the applied static magnetic field. When the magnetization is close to its stable orientation, radiation damping leads to a reduction of the lifetimes of the observed signals, known since the early days of NMR [1]. Radiation damping can also lead to unstable precession or to maser operation, and additional modifications of NMR precession (frequency shifts due to so-called cavity pulling) result from sample-resonator coupling if the latter is detuned. In this paper we will focus on the control of radiation damping, but cavity pulling can likewise be manipulated.

A high Q-factor also narrows the bandwidth $\Delta f = f_L/Q$ of the detection circuit and leads to a long characteristic response time Q/f_L for any changing signals. This introduces severe limitations in *low frequency* NMR, where recovery from equipment saturation and ring-down due to rf pulses can prevent the observation of short-lived NMR signals. It also limits the spatial resolution of low frequency magnetic resonance imaging (MRI), for which broadband acquisition of rapidly changing signals in strong imaging gradients is required.

The usual solution to reduce radiation damping consists in decreasing the Q-factor or the filling factor, i.e., degrading the reception properties of the resonator. In doing so, the SNR is unfortunately reduced as well. For MRI applications, Q-spoiling can indeed be used to address the bandwidth limitation issue as well [2], but this also reduces SNR. Another solution to the bandwidth limitation consists in correcting the recorded signal using a numerical filter in post-processing [3, 4]. This compensates for the frequency response characteristics, but strongly degrades SNR at the edges of the resonator bandwidth, and may introduce artifacts in MRI when large corrections are involved. Transiently reducing the Q-factor whenever saturating rf pulses are applied by introducing so-called Q-switches is a widely used method to limit the adverse impact of a high Q-factor on fast recovery, but this only partly alleviates ringdown problems.

Active feedback schemes have been proposed and successfully used to independently address these issues. To our best knowledge, Hoult was the first to use a negative feedback scheme together with a Q-switch device [5] to efficiently reduce recovery time for NMR operating at 5 MHz without loss of SNR. Two feedback schemes [6] have been proposed to strongly reduce current in resonators, hence to suppress radiation damping: The 'Brussels' scheme [7] uses inductive coupling of the suitably amplified signal, in such a way that the feedback emf tends to oppose the emf induced by precessing magnetization. The 'Ecole Polytechnique' scheme [8] uses directional couplers to feed back the suitably amplified signal to the NMR resonator. For MRI applications, the only non-dissipative bandwidth broadening scheme we are aware of is based on the use of over-coupled high Q NMR coils [9]. It provides adequate bandwidth for operation at several MHz with no SNR penalty.

In this paper we propose an implementation of an active feedback circuit that solves both the coupling and the bandwidth problems at very low frequency, and we report on its efficiency in two applications involving hyperpolarized ^{3}He in NMR and MRI around 2.5 mT. The device relies on the same inductive coupling of the feedback as the 'Brussels' scheme but operates at high impedance and has been used for different purposes. Beyond these examples, the potential uses of such a scheme cover all situations where non-perturbing, accurate low field NMR measurements are performed, e.g., to monitor polarization in gas hyperpolarization systems, in polarized ^{3}He targets or neutron spin filters, and in ^{3}He or ^{129}Xe magnetometry.

2. An active feedback scheme for low frequencies

Figure 1 shows the detection circuit schematic including the active feedback scheme we use for low-frequency (below 100 kHz) NMR signal detection. At such low frequencies, propagation delays in connecting cables are totally negligible and impedance matching with 50 Ω cables is useless. It is actually more convenient to use high-impedance detection tank circuits and preamplifiers having medium to high input impedance and suitably low voltage and current noise. We report results obtained using a Stanford Research Systems 560 preamplifier (voltage noise 4 nV/$\sqrt{\text{Hz}}$, current noise 2 fA/$\sqrt{\text{Hz}}$, input impedance 100 MΩ + 25 pF gain 1-50,000). Impedances of the tank circuits at resonance are 2.5 kΩ and 35 kΩ in our applications, thus the effect of current noise is negligible. Signal recordings are made from the output of this preamplifier. An electronic circuit called a "feedback control box" is used to modify the relative phase and amplitude of the signal fed back into the detection circuit. This control is essential as it determines the reduction factor of the interaction between the sample and the detection coil and ensures the stability of the closed loop circuit.

The feedback control box is built using standard low frequency electronic circuitry. We use a phase shifter [10] to change the phase of the loop signal relative to the output signal without affecting its amplitude. This decoupling between the controls of amplitude and phase allows one to easily realize optimal and reproducible settings. The electronic loop is coupled to the tank circuit using a transformer with a small self-inductance compared to the detection

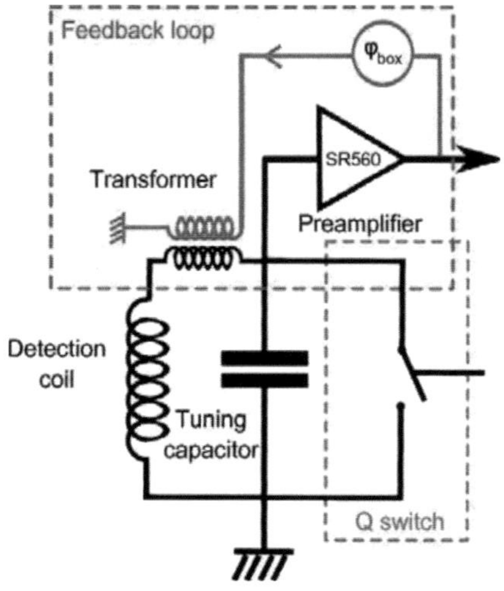

Figure 1. Active feedback schematic. The detection tank circuit, tuned at Larmor frequency f_L, is formed by the detection coil in parallel with a capacitor. It includes the secondary winding of a low-inductance transformer used with the feedback control box $\varphi_{\rm box}$ for adjustable feedback. In the MRI application, an electronic switch (Q-switch) is connected in parallel with the tank circuit. This fast-switching device allows for a strong reduction of the Q-factor as its resistance is of the order of 50 Ω when closed, much lower than the impedance of the tank circuit at resonance (35 kΩ).

coil self inductance (1/100th), thus leaving the resonance frequency of the tank circuit almost undisturbed. The transfer function of the feedback loop mainly determines the spoiling of the Q-factor and may also produce a small shift in the resonance frequency of the detection circuit, depending on the details of the feedback loop. It is convenient to distinguish the Q-factor in the open loop configuration Q, which is the one of the free tank circuit, and the Q-factor of the closed loop circuit Q_{FB}, which is the Q-factor associated with the linewidth of the frequency response of the circuit with the active feedback.

As with any active electronic circuit, the detection circuit stability depends on the choice of the feedback settings. Several features can lead to deviations from the expected behavior. In particular, one has to consider the linear Nyquist stability of the circuit. The use of the feedback control box allows for a fine tuning of the phase necessary to meet the Nyquist stability criterion with high feedback gains. The dynamic range of the feedback scheme is also limited by the non-linearity and saturation of the amplifier used in the loop.

In the next section, we describe two experiments operating at similar frequencies but for which detection circuits have different Q-factors, Johnson noise, and bandwidth: ^{3}He NMR at low temperature and ^{3}He MRI at room temperature. We thus explore two different regimes for the active feedback configuration, but in both experiments Q_{FB} of order $Q/40$ was achieved.

3. Two applications of the low frequency active feedback device

3.1. Reduction of radiation damping in low temperature ^{3}He NMR

Our group performs NMR studies on dense hyperpolarized ^{3}He-^{4}He liquid mixtures (at 1K), in which the distant dipolar field resulting from the large sample magnetization plays the dominant role in the NMR free evolution, provided that radiation damping is avoided [11]. To assess the effect of the interaction between the sample and the coil, we use ^{3}He gas at 4.2 K instead of liquid: with this reduced density, distant dipolar effects have a negligible role in NMR evolution. Hyperpolarized ^{3}He is contained in a 0.44 cm^3 spheroidal cell. NMR is performed using a shielded transmit coil orthogonal to a saddle-shaped detection coil, which is part of a tank circuit resonant at 74 kHz with $Q = 14$. This modest Q-factor results from resistive losses in the cryogenic coaxial cable used between the coil and the tuning capacitor, but in spite of this the thermal (Johnson) noise is small compared to the preamplifier noise. Signal acquisition and rf sequences, which can be quite elaborate on liquid samples [11], are performed using a Tecmag Apollo console.

For the radiation damping studies used here as an illustration of the active feedback detection performance, hyperpolarized ^{3}He gas with polarization up to 20% is used. Following a small tipping angle rf pulse, an exponentially decreasing signal is observed. Since the intrinsic transverse relaxation time is long (hours), relaxation is due to fast atomic diffusion in the weak field inhomogeneities. In this motional averaging regime, exponential relaxation is expected, with rates depending only on the field map and on the spin diffusion coefficient. However, decay rates which scale with signal amplitudes are observed. This behaviour is characteristic of the radiation damping effect for magnetization orientation near the stable direction [1, 6].

Figure 2 shows typical free induction decay signals for experiments in which different configurations of the detection scheme have been used. For the standard detection coil (a) for which the Q-factor is Q=14, the signal damping rate is 0.22 s^{-1}. When a suitable resistor

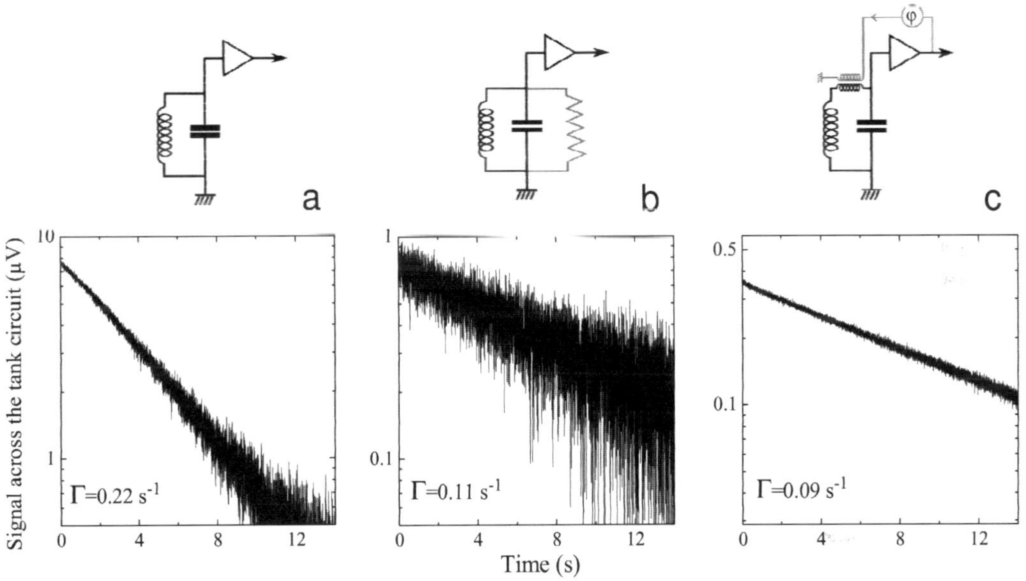

Figure 2. Signals recordings in semilog scale for three circuit configurations: standard tank circuit (a), Q-spoiled circuit (b), and tank circuit with feedback (c). These recordings have been obtained using hyperpolarized ^{3}He gas at 4.2 K (polarization ~14%, pressure ~1 torr) after a 9° tipping rf pulse. The vertical scales take into account the electronic gains and refer to voltage amplitudes across the capacitor.

is used in parallel with the capacitor to spoil the Q-factor to Q_S=1.4 (b), the signal damping rate is reduced to 0.11 s^{-1}, but the signal is attenuated (÷10) and the SNR is significantly reduced. When active feedback is used (c), a radiation damping-free signal is obtained with a damping rate of 0.09 s^{-1} and a signal to noise ratio similar to that of case (a) in spite of a smaller signal amplitude (÷22 in this case). The additional damping rate introduced by coupling to the unspoiled detection circuit is found to be $\Gamma_{\mathrm{RD}} \sim 0.13$ s^{-1}. This is consistent with expectations relying on tip angle calibrations for pulsed NMR performed using the detection coil for both rf transmit and pick-up.

The striking feature of the signals obtained using active feedback is that in spite of the strong reduction of the voltage across the detection coil (Figure 2c), the SNR is fully preserved: no passive dissipation is introduced, and the amplifier noise is reduced by the feedback loop with the same factor as the detected signal. This detection scheme is now systematically used instead of the initial Q-spoiled scheme [11] to suppress radiation damping in our NMR studies of hyperpolarized liquid helium. It will also be used to enhance the coupling between the coil and the sample magnetization in a well-controlled way. This will allow further studies of the combined effects of distant dipolar fields and radiation damping that can lead to yet another kind of spin turbulence [12].

3.2. Fast, broadband detection in ^{3}He MRI

We use a home-built low frequency MRI setup designed for small animals [13] that operates at f_L=85 kHz (corresponding to a static magnetic field of 2.7 mT). Combining Litz wire in detection coils with low-loss capacitors, Q-factors typically range from 100 to 200 depending on coil size. The required detection bandwidth corresponding to the frequency span due to the imaging gradient over the image's field of view is about 10 kHz. It could be simply achieved by spoiling the Q-factor of the detection circuit down to $Q_S \sim$10 with a resistor. But at such a low frequency in this MRI setup, the Johnson noise of the room-temperature unspoiled circuit is of the same order of magnitude or larger than the preamplifier noise, and both are the significant sources of total noise on the output. Thus, a simple spoiling of the circuit leads to a significant increase of dissipation as well as to a reduction of the signal output, both effects leading to a degradation of the SNR. The active feedback circuit added to the high-Q tank circuit allows us to dramatically increase the detection bandwidth without affecting the SNR. A comparison of the main computed properties of a simply spoiled circuit and a circuit with active feedback for parameters corresponding to the experimental conditions is shown in Fig. 3.

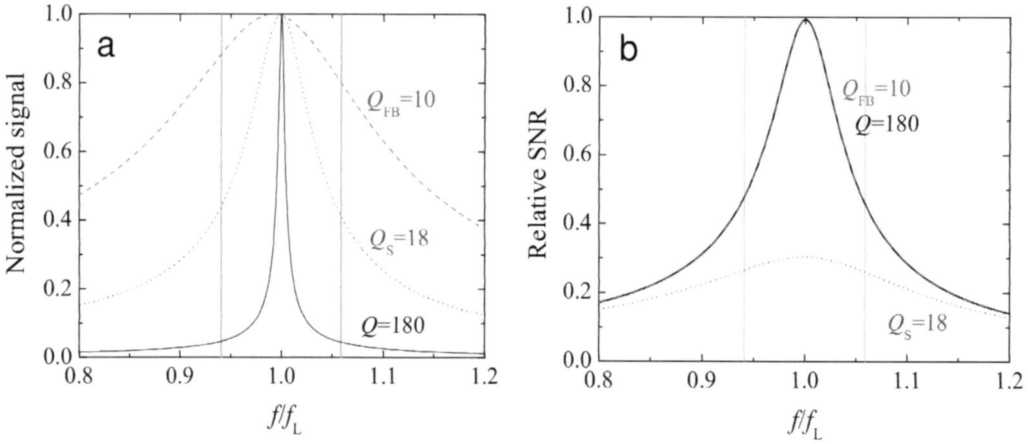

Figure 3. Computed relative amplitudes (a) and SNR (b) of the detection circuit expected for the three circuit configurations illustrated in Fig. 2. **a:** Both Q spoiling and feedback broaden the tank circuit bandwidth. For each curve, the amplitude is normalized to the value at resonance. The vertical grey lines represent the limits of the region of interest for MRI. **b:** The corresponding SNRs are normalized to the SNR of the unspoiled circuit at resonance. Feedback does not affect the SNR (overlaid upper curves) but Q spoiling decreases SNR over the MRI bandwidth by a factor of 2 to 3.

Figure 3a shows that, over the frequency range of interest, the active feedback circuit (with Q=180, Q_{FB}=10) has a flat enough frequency response and only requires small amplitude corrections, whereas the Q-spoiled and unspoiled circuits would require significant and very large corrections, respectively. Active feedback does not change the SNR (Fig. 3b): signal and noise are identically scaled by the feedback at all frequencies. The SNR of the Q-spoiled circuit is substantially degraded due to increased dissipation and to an increasing noise contribution from the preamplifier. The feedback scheme thus combines the high SNR of the high-Q circuit and the suitably flat amplitude response of a Q-spoiled circuit.

Another important advantage of using active feedback is the faster recovery of the detection circuit. Simple experimental examples of ringdown after rf-pulse saturation of a detection system with Q=100 are shown in Fig. 4. Using active feedback decreases the 'dead-time' and

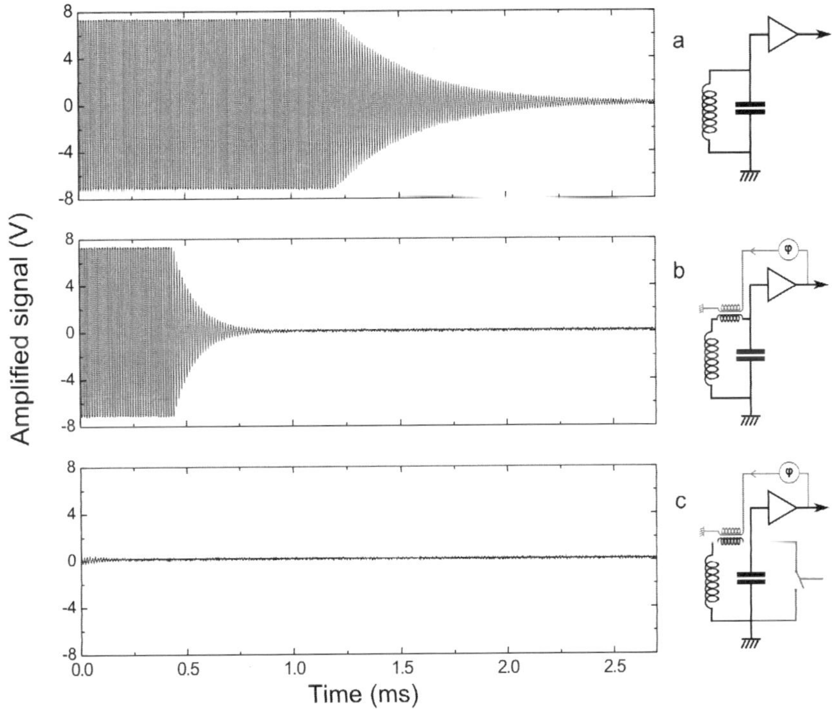

Figure 4. Experimental ringdowns recorded just after the end of a typical NMR tipping pulse. Recovery from saturation of the detection system is shown for different configurations of the detection circuit. **a**: A standard Q=100 tank circuit was used. **b**: The active feedback circuit is used, providing a reduced Q-factor Q_{FB}=25. **c**: Both active feedback and Q-switch were used. The duration of the ringdown is then sufficiently short to make acquisitions as soon as 100 μs after the end of the tipping pulse.

using an additional Q-switch (strong spoiling of the Q-factor of the coil during the rf-pulse) almost removes it completely. In the test experiment of Fig. 4c, a small parasitic signal remains for 150 μs, probably because of the Q-switch energy injected into the tank circuit during the switch opening, so data acquisition would be affected and must be blanked only during this time. This allows us to use fast pulse sequences, with acquisition periods as short as to 2 ms. The improvement is made possible thanks to the availability of low charge injection CMOS analog switch circuits which introduce less than 1 pC into the tank circuit when switching, thus minimizing disturbance of the NMR acquisition.

Figure 5 is an illustration of a successful utilization of active feedback in a low field MRI experiment. The active feedback has been used to obtain an image of gas in a sealed glass cell (1.8 cm × 12 cm) containing 400 mbar of ^{3}He polarized to a fraction of a percent[1] using a Fast Low Angle SHot (FLASH) sequence. Short acquisition times had to be used to avoid strong signal attenuation in highly diffusive gas, so echoes were recorded for 4 ms with a 10 kHz sampling rate. The tank circuit used for this image consisted of a pair of rectangular coils

[1] At high pressure, laser polarization is quite inefficient which explains this very small polarization. Still, total magnetization is higher than in typical low-pressure cells, and slower gas diffusion makes MRI possible

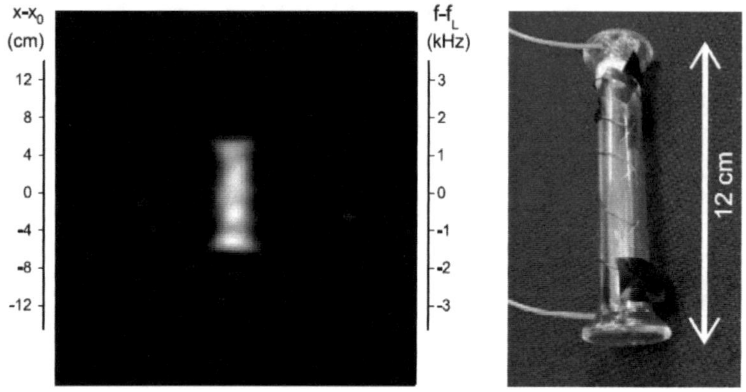

Figure 5. MRI image (left) obtained on a sealed glass cell (pictured on right) using hyperpolarized ^{3}He (polarization $\sim$ 0.5%). A FLASH imaging sequence was used (32 $\times$ 38 points, 12° tip angle, echo time 2.5 ms, acquisition time 4 ms).

(5 cm$\times$15 cm, 4 cm apart, 36 turns each, Litz wire consisting of 25 strands of 0.1 mm) and a low-loss C0G capacitor (2.2 nF). Its high Q-factor (Q=180) would normally yield a grossly insufficient bandwidth (f_L/Q=0.5 kHz) but the use of feedback extended the detection bandwidth to about 6 kHz. As a result, the noise pattern is quite uniform over the image, a feature made possible by the use of the active feedback instead of bandwidth correction as was done in references [3, 4]. Additionally, the scheme lessens ring-down effects that plague low field NMR. This could be useful for solid state NMR at low frequencies.

Acknowledgments

This research program is supported in part by the Agence Nationale de la Recherche (ANR DIPOL grant) and by the European Commission (PHeLINet Marie Curie Research and Training Network grant).

References

[1] Bloembergen N and Pound R V 1954 *Phys. Rev.* **95** 8
[2] Tsai L L, Mair R W, Rosen M S, Patz S and Walsworth R L 2008 *J. Magn. Reson.* **193** 274–285
[3] Bidinosti C P, Choukeife J, Nacher P J and Tastevin G 2003 *J. Magn. Reson.* **162** 122 – 132
[4] Ruset I C, Tsai L L, Mair R W, Patz S, Hrovat M I, Rosen M S, Muradian I, Ng J, Topulos G P, Butler J P, Walsworth R L and Hersman F W 2006 *Concepts Magn. Reson. B* **29B** 210–221 ISSN 1552-5031
[5] Hoult D I 1979 *Rev. Sci. Instrum.* **50** 193–200 URL `http://link.aip.org/link/?RSI/50/193/1`
[6] Jeener J 2002 *Encyclopedia of Nuclear Magnetic Resonance, Volume 9: Advances in NMR* (John Wiley & Sons)
[7] Broekaert P and Jeener J 1995 *J. Magn. Reson. A* **113** 60–64
[8] Louis-Joseph A, Abergel D and Lallemand J Y 1995 *J. Biomol. NMR* **5** 212–216
[9] Raad A and Darrasse L 1992 *Magnetic Resonance Imaging* **10** 55 – 65 ISSN 0730-725X URL `http://www.sciencedirect.com/science/article/B6T9D-4BX2CCT-8/2/82914e7fb2ebf60c653330b5800967fb`
[10] Horowitz P and Hill W 1989 *The Art of Electronics* (Cambridge University Press)
[11] Hayden M E, Baudin E, Tastevin G and Nacher P J 2007 *Phys. Rev. Lett.* **99** 137602
[12] Lin Y Y, Lisitza N, Ahn S and Warren W 2000 *Science* **290** 118–121
[13] Nacher P J, Pelissier M and Tastevin G 2007 *Proc. Intl. Soc. Mag. Reson. Med. 15* p 3290 ISSN 1545-1428

Journal of Physics: Conference Series **294** (2011) 012010 doi:10.1088/1742-6596/294/1/012010

Low magnetic field manipulation of ^{3}He spins using digital methods

S R Parnell[1,3*], S Boag[2], T J McKetterick[2], J M Wild[1]
[1]Academic Unit of Radiology, University of Sheffield, S10 2JF, UK
[2]ISIS Neutron and Muon Facility, Rutherford Appleton Laboratory, Chilton, Didcot, Oxfordshire, OX11 0QX, UK
E-mail: stevparn@indiana.edu

Abstract. Simple methods for manipulating ^{3}He spins at low magnetic fields using low cost digital I/O cards are presented. A brief overview of a digital NMR spectrometer is given along with practical information on the implementation and an assessment of the losses per pulse. Through the use of coils with different quality factors the effects of radiation damping are investigated and assessed in the context of the sample polarisation using neutron transmission measurements. The hardware is further used to synthesize an adiabatic swept pulse which is used to reverse the ^{3}He spin direction, allowing the selection of a particular neutron spin.

1. Introduction

Spin-polarised or 'hyperpolarised' (HP) noble gases are now routinely used across a range of disciplines such as neutron physics [1], polarised targets [2] materials science [3] and medical imaging [4]. The high level of polarisation is usually created by using either metastable [5, 6] or spin exchange [7] optical pumping. The optical pumping is typically performed in a low magnetic field, $\sim$ 10-30G, which defines the optically pumped state and minimises polarisation losses due to magnetic gradients. These fields correspond to Larmor frequencies of the order $\sim$ 30-100kHz (for ^{3}He), which can now be easily synthesized using low cost digital I/O cards allowing filtering, RF synthesis and timing to be easily controlled using a minimal amount of hardware. Due to the high nuclear polarisations which are now routinely obtainable [8, 9], it is possible to detect the magnetisation of the gas with no additional pre-amplification.

2. Polarimetry by NMR

The standard method of measuring the relative polarisation of HP ^{3}He is by pulse acquire NMR which measures the free induction decay (FID) signal from the excited nuclear spins. For most measurements it is important that the fraction of spins destroyed by each measurement is small in comparison to the total magnetisation. Small surface coils on the cell walls are used as standard. Accurate measurement of the polarisation is necessary in many different applications of hyperpolarised gases. For example in studies into polarisation limits [9] where the X-factor determination requires accurate knowledge of the polarisation build up or 'spin

[3] present address: Center for Exploration of Energy and Matter University of Indiana, Bloomington, IN 47408 USA

up' and relaxation (T_1) time constants. Or neutron scattering where knowledge of the ^{3}He polarisation is used to calculate the neutron polarisation and hence correct for the polariser efficiency.

For data analysis the shape of the FID is well described by equation 1;

$$f(t) = \frac{1}{2} V_{pp} g(t) sin(\omega t + \phi) \tag{1}$$

where V_{pp} is the peak-to-peak voltage a t=0 with ω the difference between the Larmor frequency and the transmit frequency of the pulse, g(t) the envelope function to describe the FID lineshape and ϕ an arbitrary phase angle to account for the phase shift between transmit and receive pulse.

2.1. Hardware

The hardware required for implementing a simple digital NMR spectrometer consists of a PC with a digital I/O card (e.g. NI-6251, 1.25MHz), a BNC connection box, a home made T-R switch (duplexer)[10] and a combined transmit/receive (T_x/R_x) coil, which is grounded to the aluminium frame of the coils. In addition a variable capacitance box (100pF to 11μF) is used to adjust the tuning of the coil such that the resonance frequency can be easily changed for use at different magnetic field strengths. In-situ polarimetry of gas in the cell typically relies on small tip angle NMR from a coil close to the cell. The transmit/receive coil is 3cm in diameter, hand wound from 0.25mm wire with ~300 turns and has a quality factor Q of 10. For higher values of Q the coil is spoilt using a resistor of a few ohms. Typical ring-down times for the coil are of the order 6ms. It is important to have a low Q coil as distortion of the FID can occur due to the high level of magnetisation (see section 2.3).

The simplicity of the hardware is due to the passive T-R switch, which consists of 4 (high speed 1N4148) diodes and a resistor. This enables the use of the single T_x/R_x coil to measure longitudinal (T_1) and transverse (T_2^*) relaxation time [11] in different regions. The spectrometer has been used to asses and optimise polarisation losses in end compensated solenoids, mu-metal boxes [12] and Helmholtz coils. Such optimisation is performed by minimising the NMR linewidth i.e. achieving a long T_2^*. However, care must be taken to sample enough of the volume of the cell as the small diameter coil only samples a small region. To overcome this limitation it is possible to use this system with larger coils, which encompass the whole cell volume and inevitably the whole cell T_2^* is shorter due to the B_0 inhomogeneity. However, this introduces a problem as the larger coils will tend to destroy more polarisation per excitation pulse.

2.2. Losses

In order to accurately measure the spin up and relaxation (T_1) constants it is essential that the destruction per NMR pulse is negligible or well characterised. In order to check this we performed a series of 500 pulses sequentially with a total measurement time of ~25 minutes. As the T_1 of the cell under investigation was 90 hrs (measured via neutrons[13]) then any contribution from the normal decay is negligible and all losses are assigned to the NMR. Shown in figure 1 is the initial amplitude of each pulse (Vpp). The total destruction per pulse is 1×10^{-4} per pulse. Figure 1 inset shows a typical pump up of a cell with a spin up constant of 849$\pm$ 8.8 minutes. If we correct the data for the pulse destruction losses and fit to this data we get an increase of less than one part in ten thousand, therefore these losses are accepted as negligible.

2.3. Radiation Damping

It is important to note the effects of the high magnetisation on the observed FID from the optical pumping cell in both the high and low energy states and the effect of radiation damping on the observed total magnetisation. Radiation damping [14] occurs when spins are tipped from

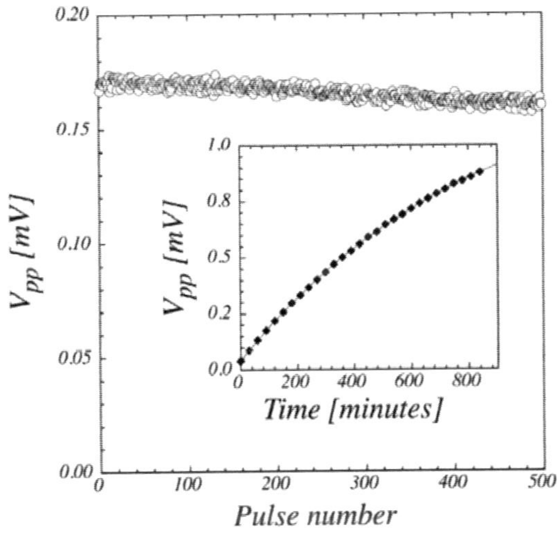

Figure 1. Destruction per pulse. Inset : typical spin up data uncorrected for polarisation losses.

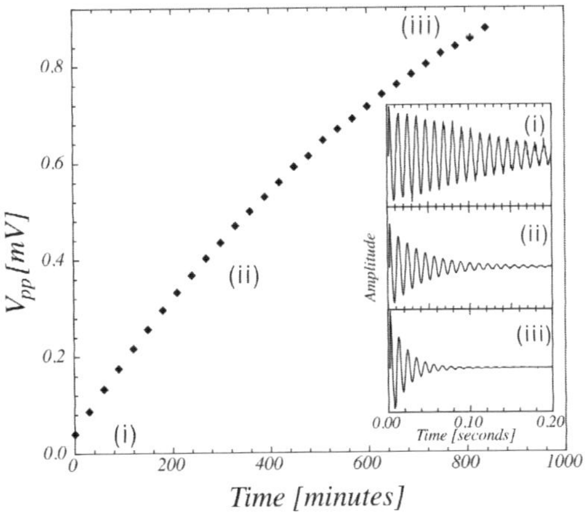

Figure 2. Spin up curve for the high Q coil with the peak voltage determined by fitting of the FID to equation 1. Inset :FID's taken at various levels of polarisation, indicated (i),(ii) and (iii).

a longitudinal to a transverse plane. For sufficiently high magnetisation the precessing spins create a 'counter field' in the receive coil which acts to re-orientate the magnetisation back into the longitudinal plane. This has two principle effects upon the measurement. The first is that the effective tip angle experienced depends upon the sample magnetisation and T_2^*. Secondly the measured FID T_2^* or linewidth (in the Fourier domain) will change with sample magnetisation.

Radiation damping [15] changes the shape of the FID by addition of a term $\text{sech}(t/\tau_{rd})$ to the original T_2^* decay where τ_{rd} is the radiation damping time constant and therefore is observed for long T_2^* values. τ_{rd} is given in SI units by;

$$\tau_{rd}^{-1} = Q\gamma m_0 \eta \mu_0 / 2 \qquad (2)$$

where Q, γ, m_0, η are the coil quality factor, gyromagnetic ratio, longitudinal magnetisation density and filling factor respectively [16]. Radiation damping from ^{3}He gases has been observed in a number of different studies. For example MRI studies at low field (20.6 G) were performed by Wong et al [17]. These identified radiation damping by using a constant RF pulse and different tuning of the receive coil. They noted a difference in the calculated tip angles indicative of radiation damping. At higher fields (1.5T) effects have also been observed [18] using highly polarised samples in birdcage resonators

In order to investigate these effects the polarisation build up was monitored using the spectrometer detailed earlier with two orthogonal T_x/R_x coils of different Q. For low Q and m_0 the radiation damping should be negligible however at higher values radiation damping effects can occur, these will be manifested in a narrowing of the FID linewidth. In order to observe this the two coils of differing Q were mounted on the same optical pumping cell, hence measuring the same m_0. The coils were in a direction mutually orthogonal to the B_0. The cell used in this investigation was a 2.0Bar cell with a saturation polarisation of 69% [13], polarised in the low energy state by spin exchange optical pumping using a 40W narrowed external cavity diode laser (ECDL) [19].

The different Q values were obtained by winding different number of turns (100 Q=35 and 300 Q=3), in addition the 300 turn coil had a 10Ω resistor in series in order to artificially dampen the Q. Q was measured using a function generator and search coil to transmit and the pickup was measured using an oscilloscope connected to the coil. All experiments were carried out in a B_0 field of $\sim$10G using a Helmholtz pair (diameter 0.8m) with a uniform magnetic field over the cell volume. Pulses were 0.7V with a duration of 1.2ms and in each case the coils were pulsed sequentially with a delay of $\sim$1s between them, therefore over the course of a spin up (12 hours) they are measuring the same magnetisation (within error).

Shown in figure 2 is the initial FID voltage (V_{pp}) as a function of time whilst a 2 Bar ^{3}He cell is polarised. This is measured using the high Q coil, the inset shows the recorded FID's. clearly observable is a shortening in the decay time with increasing magnetisation (polarisation).

Figure 3 shows the change in linewidth with increasing polarisation (spin- up time) for the two coils. The low Q coil has a smaller change, during the start of the measurement at low m_0 both coils have a similar observed linewidth and converge to a value $\ll 200ms$ representing the background T_{2*}, however with increasing polarisation the higher Q coil has an observed T_{2obs} which changes more significantly. Analysis of spin up time gives two different times $(Q = 35)849 \pm 9min$ and $(Q = 3)722 \pm 26min$. This is consistent with the high Q coil having a lower effective tip angle as expected due to the radiation damping and shows that any polarisation measurements inferred from NMR with this coil are more sensitive to error, especially at high polarisation.

Absolute measurements of the ^{3}He polarisation were performed on the neutron beamlines POLTAX and CRISP[13, 20] whilst simultaneously measuring the polarisation via NMR. Typical NMR data for two different coils is shown in figure 4. As the neutron transmission measurement is an absorption measurement it is not effected by radiation damping effects and hence is an accurate measurement of the gas polarisation. The data was collected using a two coil system similar to the lab system described earlier. A polarised ^{3}He cell of 2.16 Bar was placed in a 'magic box' [12] and allowed to decay whilst simultaneously measuring the relative gas polarisation using NMR and absolute polarisation using neutrons. The neutron data was measured as a time series and interpolated to match the NMR data. In these measurements good agreement was obtained between the NMR and neutron data for coils with Q of up 10, as indicated by the linear fit to the data . From equation 2 it can be seen that the total 'figure of merit' $(Q \times pressure \times polarisation)$ is at least as high as $\sim$ 12. However this is only valid for monitoring the ^{3}He polarisation in the low energy state.

Shown in figure 5 are a series of FIDs when pumping in the high energy state, this was done by adjusting the circular polarisation of the pump beam (relative direction of the B_0 and $\lambda/4$

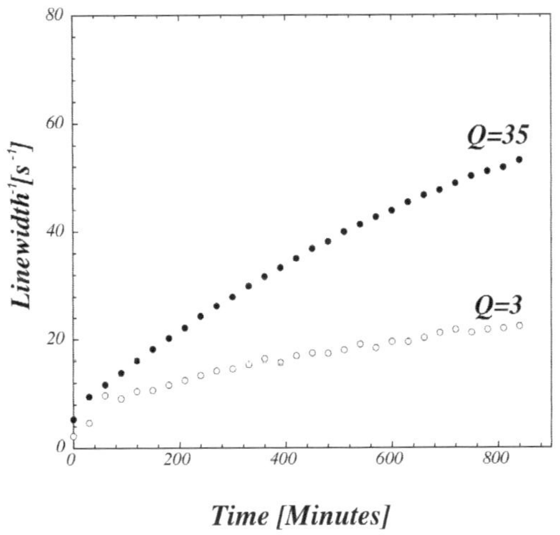

Figure 3. Linewidth^{-1} from exponential fit for the two different coils during polarisation build up.

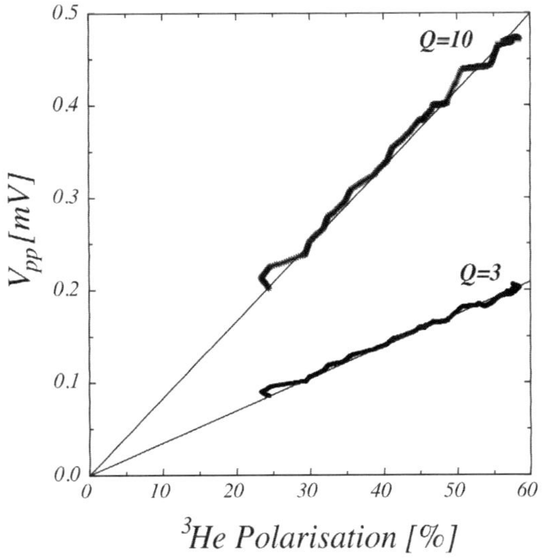

Figure 4. NMR peak voltage for two different coils as a function of ^{3}He polarisation measured by neutron transmission. Line is to indicate the linearity between the neutron and NMR measurements

polarising plate). The data is labelled (i), (ii) and (iii) in increasing polarisation. The form of the decay is non-exponential obeying a simple sech relationship and can be used to determine when pumping in the high energy state, however no reliable NMR polarimetry can be performed with these FID's for most cases it is preferable to reverse the spin state using adiabatic fast passage (AFP) [11] methods.

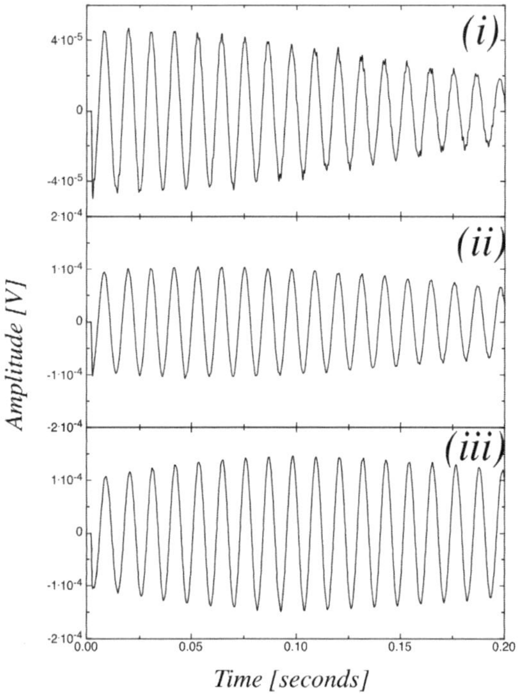

Figure 5. Typical FID data obtained from ^3He pumped into the high energy state. Polarisation (magnetisation) increasing with number.

3. ^3He spin reversal

AFP methods have been used in conventional NMR for decades [11], however recently these techniques have been applied to neutron spin filters on neutron beams allowing the construction of a combined neutron polariser and flipper [21, 22]. These rely on the use of an effective magnetic field which the nuclear spins follow from an alignment parallel (anti-parallel) to the holding field B_0 to an alignment anti-parallel (parallel).

Practically the RF field waveform is a sine wave which is swept through the Larmor frequency of the ^3He, in addition to this the function is convoluted with a Gaussian envelope. This envelope serves to avoid abrupt changes in the total magnetic field which may give rise to non adiabatic transitions and lead to depolarisation. This results in a function of the form;

$$B_{RF}(t) \propto exp\left(-\frac{[\omega(t)-\omega_L]^2}{\sigma^2}\right) sin\left[2\pi\omega(t)\times t\right] \tag{3}$$

where ω_L is the resonant frequency and σ is the width of the Gaussian envelope. Earlier systems [22] employed dedicated analogue electronics for the synthesis of the swept waveform, in this work the polarisation losses were 2 x 10^{-5} per AFP flip. These losses were attributed to noise in the AFP electronics. In a more recent work the same digital I/O card used for NMR in the earlier section was adapted to provide digital synthesis of the RF AFP waveform[23]. With an additional analogue amplifier losses of $(1.78\pm0.03)\times10^{-5}$ per AFP flip were obtained. Moreover the digital system requires minimal hardware and can be easily reproduced allowing routine spin selection of the neutron polarisation with a negligible effect on the ^3He gas polarisation.

Combining this on a neutron beam line with the NMR allows for a compact polarimetry system which is capable of determining the polarisation and T_1 of the gas (via NMR), which is necessary for polarisation corrections along with rotation of the polarisation (via AFP) and monitoring the magnetisation state of the gas (via FID lineshape).

4. Discussion

The work presented in this paper illustrates how simple instrumentation for polarisation manipulation and measurement of ^{3}He can be produced. Such systems can also be used for other noble gases such as ^{129}Xe [24], however for low ^{129}Xe concentrations averaging is required to achieve a high signal to noise ratio. Furthermore the modularity of these digital systems allows additional outputs to be synchronised with the T_x pulse allowing the addition of gradients (with suitable amplifier and coils) to the system which can be utilised for the measurement of cells pressure, temperature and polarisation distribution [25].

5. Acknowledgments

This research project has been supported by EPSRC [EP/D070252/1], STFC [CC/D000130/1] and the European Commission under the 6th Framework Programme through the Key Action: Strengthening the European Research Area, Research Infrastructures. Contract no: RII3-CT-2003-505925

6. References

[1] Gentile T, Babcock E, Borchers J, Chen W, Hussey D, Jones G, Lee W, Majkzrak C, O'Donovan K, Snow W, Tong X, Velthuis S, Walker T and Yan H 2005 *Physica B: Condensed Matter* **356** 96-102
[2] Zheng X, Aniol K, Armstrong D S, Averett T D, Bertozzi W, Binet S, Burtin E, Busato E, Butuceanu C, Calarco J, Camsonne A, Cates G D, Chai Z, Chen J P, Choi S, Chudakov E, Cusanno F, De Leo R, Deur A, Dieterich S, Dutta D, Finn J M, Frullani S, Gao H, Gao J, Garibaldi F and Gilad S 2004 *Phys. Rev. C* **70** 065207
[3] Goodson B M 2002 *J. of Magnetic Resonance* **155** 157-216
[4] Otten E W 2004 *Europhysics News* **35** 16
[5] Colegrove F D, Schearer L D and Walters G K 1963 *Phys. Rev.* **132** 2561-72
[6] Parnell S R, Wallidge G W, Frost C D, Langridge S, Eccleston R S and Phillips P 2005 *Physica B: Condensed Matter* **356** 114-7
[7] Bouchiat M A, Carver T R and Varnum C M 1960 *Phys. Rev. Lett.* **5** 373-5
[8] Parnell S R, Babcock E, Nünighoff K, Skoda M W A, Boag S, Masalovich S, Chen W C, Georgii R, Wild J M and Frost C D 2009 *Nuclear Instruments and Methods A* **598** 774-8
[9] Babcock E, Chann B, Walker T, Chen W and Gentile T 2006 *Phys. Rev. Lett.* **96** 083003
[10] Parnell S R, Woolley E B, Boag S and Frost C D 2008 *Measurement Science and Technology* **19** 045601
[11] Abragam A 1961 *Principles of Nuclear Magnetism* (Oxford University Press)
[12] Petoukhov A, Andersen K, Jullien D, Babcock E, Chastagnier J, Chung R, Humblot H, Lelievre-Berna E, Tasset F, Radu F, Wolff M and Zabel H 2006 *Physica B: Condensed Matter* **385-386** 1146-8
[13] Boag S, Parnell S, Frost C, Andersen K and Babcock E 2007 *Physica B: Condensed Matter* **397** 179-81
[14] Bloembergen N and Pound R V 1954 *Phys. Rev.* **95** 8
[15] Augustine M P 2002 *Progress in Nuclear Magnetic Resonance Spectroscopy* **40** 111-50
[16] Gueron M and Leroy J L 1989 *J. of Magnetic Resonance* **85** 209-15
[17] Wong G P, Tseng C H, Pomeroy V R, Mair R W, Hinton D P, Hoffmann D, Stoner R E, Hersman F W, Cory D G and Walsworth R L 1999 *J. of Magnetic Resonance* **141** 217-27
[18] Teh K, de Zanche N and Wild J M 2007 *J. of Magnetic Resonance* **185** 164-72
[19] Chann B, Nelson I and Walker T G 2000 *Opt. Lett.* **25** 1352-4
[20] Boag S, Babcock E, Andersen K H, Becker M, Charlton T R, Chen W C, Dalgliesh R M, Elmore S D, Frost C D, Gentile T R, Anton R L, Parnell S R, Petoukhov A K, Skoda M W and Soldner T 2009 *Physica B: Condensed Matter* **404** 2659-62
[21] Jones G, Dias F, Collett B, Chen W, Gentile T, Piccoli P, Miller M, Schultz A, Yan H, Tong X, Snow W, Lee W, Hoffmann C and Thomison J 2006 *Physica B: Condensed Matter* **385-386** 1131-3
[22] Babcock E, Petoukhov A, Chastagnier J, Jullien D, Lelievre-Berna E, Andersen K H, Georgii R, Masalovich S, Boag S, Frost C D and Parnell S R 2007 *Physica B: Condensed Matter* **397** 172-5
[23] McKetterick T, Boag S, Stewart J, Frost C, Skoda M, Parnell S and Babcock E 2010 *Physica B: Condensed Matter* In Press, Corrected Proof, DOI: 10.1016/j.physb.2010.11.088

[24] Parnell S R, Deppe M H, Parra-Robles J and Wild J M 2010 *J. of Applied Physics* **108** 064908
[25] Parnell S R, Deppe M H, Ajraoui S, Parra-Robles J, Boag S and Wild J M 2010 *J. of Appl. Phys.* **107** 094904

Journal of Physics: Conference Series **294** (2011) 012011 doi:10.1088/1742-6596/294/1/012011

Effects of high intensity neutron flux on in-situ spin-exchange optical pumping of ^{3}He

E Babcock[1*], S Boag[2], C Beecham[2], T E Chupp[3], T R Gentile[4], G L Jones[5], A K Petukhov[6] and T G Walker[7]

[1]Jülich Centre for Neutron Science, Institut für Festkörperforshung, Forschungszentrum Jülich GmbH, 85747 Garching, Germany
[2] ISIS, Chilton, Didcot OX11 QX United Kingdom
[3] University of Michigan, Ann Arbour MI 48109 USA
[4] National Institute of Standards and Technology, Gaithersburg MD 20899 USA
[5] Hamilton College, Clinton NY 13323 USA
[6] Institut Laue Langevin, 38042 Grenoble France
[7] University of Wisconsin Madison, Madison WI 55106 USA

E-mail: e.babcock@fz-juelich.de

Abstract. Prior studies revealed neutron beam-induced alkali-metal spin relaxation in ^{3}He cells polarized in-situ by spin-exchange optical pumping. These effects are minor for neutron beams of low to moderate intensity, but become important at the highest neutron flux levels available in the full spectrum beams of high flux neutron sources. It was found that the relaxation consisted of a fast and a slow component, but the origin of neither is understood. This work further explored the mechanisms of this effect by measuring the magnitude and time dependence of alkali-metal spin-relaxation rates as a function of N_2 and ^{3}He composition. The experiments were performed on a thermal neutron beam line at the Institut Laue-Langevin with a maximum neutron flux of 2.4×10^9 cm^{-2} s^{-1}.

1. Introduction

Polarized ^{3}He has applications in many areas of neutron science as a neutron polarizer or polarization analyzer as neutron spin filter cells (NSF) [1, 2]. While typically the metastability exchange optical pumping method (MEOP) [3] can produce more polarized ^{3}He, up to 1 bar-liter to 2 bar-liters per hour [1], current methods with spin-exchange optical pumping (SEOP) can also provide high production rates. Methods such as frequency narrowed diode array bars [4, 5] and hybrid SEOP [6] have led to increased performance and speed for this method such that cells containing more than 1 bar-liter to 2 bar-liters of gas can be polarized in 1 day time scales [7, 8]. Additionally both methods have achieved ^{3}He polarizations of 80% with neutron measurements [1, 9, 10]. However the ^{3}He is often polarized in a laboratory away from the neutron instrument and thus undergoes polarization decay with typical time constants on neutron instruments of 100 hours to 300 hours.

In-situ optical pumping is one way to counteract the effects of the polarization decay of the ^{3}He and potentially improve time averaged neutron performance and ^{3}He polarization. True in-situ optical pumping to maintain steady ^{3}He polarization can currently only be done with the SEOP method as it can be performed at the ^{3}He pressures required for use as a neutron spin

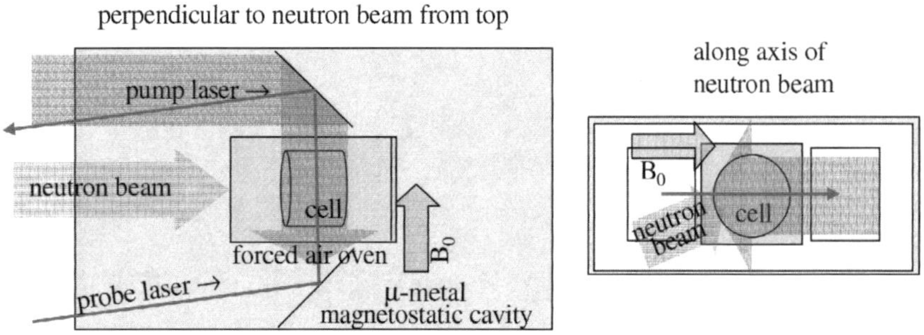

Figure 1. The spin-exchange optical pumping setup used for this experiment. One pump laser was used which counter propagated with the probe laser.

filter. However previous studies performed by members of this team, on the fundamental neutron physics beam line PF1B at the Institut Laue-Langevin (ILL)[11], had shown adverse effects to the performance of the optical pumping process in conditions of extreme levels of neutron flux which are currently normally only experienced in cold neutron particle physics beams. With increased instrument neutron flux due to ever improving neutron optics and higher intensity neutron sources, the effects could eventually be relevant to developing ^{3}He spin filters for such new instrumentation.

The reduced optical pumping performance is caused by increased alkali-metal relaxation rates which in turn increases the amount of required laser power. In extreme cases this leads to lowered alkali-metal polarization and thus lowered ^{3}He polarization. In high flux conditions this increase in alkali metal relaxation was measured vs neutron flux from 4×10^7 s^{-1} cm^{-2} to 5×10^9 s^{-1} cm^{-2} for a cold neutron spectrum (i.e. mean neutron wavelength 0.45 nm) and appeared to follow a square root of flux dependence [12]. The exact cause of the effect was not understood and it is further complicated by the fact that the effect was found not only to depend on neutron flux, but it also appeared to have two relevant time scales, one affecting the alkali metal relaxation quickly, in less than a second, and the other affecting it on time scales on the order of hundreds of seconds. Additionally, the alkali-metal relaxation seemed to depend on the amount of nitrogen buffer gas in the cell. Nitrogen, which must be present in SEOP cells to prevent radiation trapping, while probably preventing effects of ionization on direct ^{3}He nuclear spin relaxation as was observed in particle beam experiments [13, 14], seemed to be increasing the magnitude of the alkali-metal spin-relaxation for the neutron beam case. Consequently, additional measurements were conducted in order to explore the the nature of the two time constants observed in the first experiments and also the nitrogen pressure dependence of the effect.

2. Experiment

The same apparatus as in Ref. [15] was installed on the tomography beam line at the ILL and is shown in Fig. 1. This beam employs thermal neutrons with an assumed mean neutron wavelength of 0.18 nm and a particle flux of 2.4×10^9 s^{-1} cm^{-2}. The flux was calibrated using the gold foil activation method and was a constant intensity over the duration of the experiment. The neutron flux could not be varied in a calibrated way to do flux dependent tests. Since this beam has a different mean neutron wavelength than in [12] it has a different fractional neutron absorption for a given cell and has roughly half the absolute particle flux of the beam used in that prior work. For confirmation, the cell Lucky Luke was used in both experiments and gave a reasonable agreement in the measured effect when corrections for the differences in the two beams are made, despite the very different characteristics.

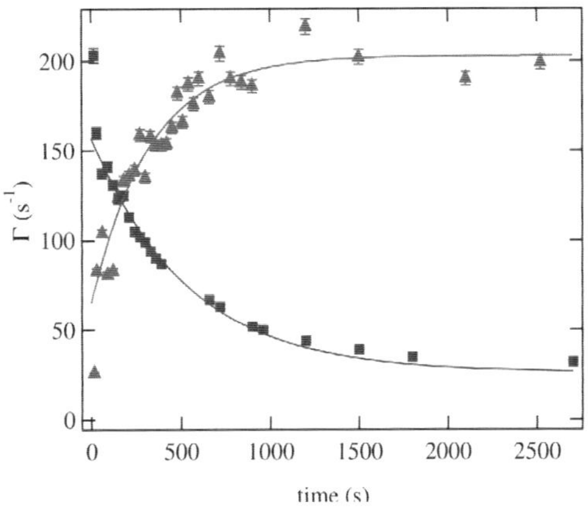

Figure 2. A sample of the measured alkali-metal relaxation rates, Γ, as a function of time after the beam was opened, red triangles, or closed, blue squares, for the cell Tin Man.

The alkali-metal relaxation was measured in the same way as in Ref. [12] using a method referred to as relaxation in the dark [16]. A diagram of the apparatus is shown in Fig. 1. The alkali-metal was polarized to low polarization by the optical pumping laser running at low power and at a wavelength a few nanometers away from the Rb resonance. A low power linearly polarized probe laser tuned to 778 nm was then used to observe the Faraday rotation, which is proportional to the Rb polarization for constant Rb density, while the pump laser was chopped at about 1 Hz to observe the decay transients of the Rb polarization. The alkali-metal relaxation was in this way measured as a function of time after the beam was opened or closed. In this paper we will always report the measured alkali-metal relaxation rates, which are a factor of the slowing down factor lower than the actual electronic relaxation rates which are proportional to the photon demand to polarize the vapor [17]. A sample of data is shown in Fig. 2. In this way we were able to extract the fraction of the effect that happened quickly at t=0, and the fraction that happened with a long time constant from the t=0 intercepts of exponential fits to the data.

The tomography beam had a "fast shutter" that could be used to block the neutron beam (but not the epi-thermal components of the spectrum which are not absorbed by the ^{3}He cell anyway) on the time scale of less than one second. From observing the response of the Faraday rotation signal while this shutter was pulsed, it appeared that the quick change in the alkali-metal polarization was coincident with the opening and closing of this shutter, verifying previous observations [12], but with more confidence.

For these measurements 4 cells were used, 3 permanently sealed cells of different ^{3}He and N_2 gas compositions and one valved cell that allowed us to vary the ^{3}He and N_2 partial pressures. The parameters of the cells and the different fillings for the valved cell are given in Table 1. Most of the relaxation measurement data was taken at a temperature of 170 °C as measured by a thermocouple placed near the cell. For the Lucky Luke cell data was also taken at 150 °C and 190 °C. The 190 °C data had a higher zero flux alkali-metal relaxation in agreement with the known temperature dependence of the relaxation cross sections for the gas composition of the cell. The change in measured alkali-metal relaxation showed the same magnitude and it had the same fraction of slow and fast components as the 170 °C data within uncertainties. The 150 °C data is more complex. The measured alkali-metal relaxation when exposed to the neutron flux for the first time rose from 15 s^{-1} to 126 s^{-1}, in rough agreement with the 170 °C and 190 °C data for the total change in alkali-metal relaxation. The initial zero flux relaxation rate of 15 s^{-1} before exposure to the neutron flux was again in agreement with the known cross sections

cell name	[^{3}He](bar)	[N_2](bar)	L_1 (cm)	L_3 (cm)	A_n	$\Delta\Gamma$ (1/s)±5%
Lucky Luke	0.5	0.1	5.0	1.6	0.33	100
Tin Man	0.5	0.4	2.0	0.6	0.28	176
Scarecrow	0.5	0.02*	8.4	3.0	0.28	36
valved cell	0.64	0.12	4.0	1.3	0.34	141
	0.90	0.02	5.1	1.8	0.45	119
	2.4	0.05	1.9	0.7	0.79	296
	2.2	0.4	1.2	0.4	0.77	354
	0	2.0	0.5	0.1	0	0
	1.4	0	3.6	1.4	0.60	96
	0.4	0	12.8	4.8	0.23	58

Table 1. List of cells and their parameters used in this experiment. All the cells were 5 cm by 5 cm cylinders except for Lucky Luke which is a 6 cm by 6 cm cylinder. The * on the [N_2] value is to denote this parameter is not well known due to the factors at the time of the cell filling. All of the cells contained Rb only for alkali-metal. L_1 is the stopping length of proton created by a neutron absorption and L_3 is the triton stopping length. A_n is the fraction of absorbed neutron flux assuming a thermal Boltzmann distribution with 0.18 nm peak wavelength. $\Delta\Gamma$ is the total change in measured alkali-metal relaxation. All quantities are assumed accurate to the last digit except for $\Delta\Gamma$ which has a relative standard uncertainty of 5 %. The error bars are determined from the typical uncertainties in the exponential fits to the alkali-metal relaxation vs time such as those shown in Fig. 2.

for the cells gas composition and temperature. However after 2000 s with the beam off, the new zero flux relaxation remained higher than expected at 44 s^{-1}.

The time constants for the build up and decay are not necessarily the same. As can be seen in Fig. 2 the time constant for the build up of alkali-metal relaxation after the neutron beam is opened appears to be faster than the decay after the neutron beam is closed. For the data shown in Fig. 2 the time constant was 320 s $\pm$ 50 s for the build up and 550 s $\pm$ 25 s for the decay of neutron induced alkali-metal relaxation. For the valved cell without N_2 the build up of relaxation happened very quickly, reaching its maximum value after perhaps 30 s with the decay having a time constant of 165 s $\pm$ 45 s and 320 s $\pm$ 20 s for the 0.4 bar and 1.4 bar ^{3}He data respectively. Because of this it is more difficult to extract the slow and fast components for the valved cell without N_2 for the build up data. The extracted values of the slow and fast components for the zero N_2 data points are from the decay only which gave reasonable exponential fits. For the other data points the values are the average values obtained from the zero-crossings of the exponential fits to the build up and decay. The relative standard uncertainty for this analysis is 10% of $\Delta\Gamma$ for both the slow and fast component.

The valved cell always had higher alkali-metal relaxation than the sealed cells. Even in the absence of neutron flux the measured alkali-metal relaxation rates were higher than what one would expect from the known relaxation cross sections for the filled gas composition and temperature. This was most likely due to the presence of additional gas impurities from the filling of the cell. This makes it difficult to compare the valved cell directly in absolute terms with the sealed cells. Since the measured alkali-metal relaxation without neutron flux was stable over the course of the experiments and was repeatable, we believe the data can still provide additional qualitative information on the observed effects. However as can be seen in Table 1 it appears that the magnitude of the neutron induced alkali-metal relaxation is higher for cell fillings similar to those of the sealed cells measured, therefore it is likely the unknown impurities are also influencing the neutron induced alkali-metal relaxation mechanisms.

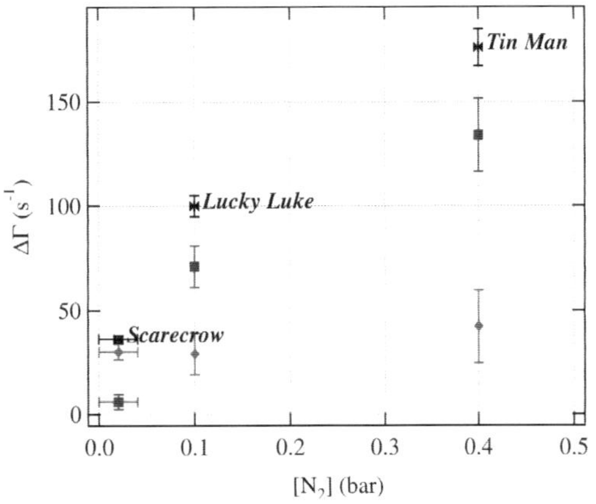

Figure 3. The measured alkali-metal relaxation versus $[N_2]$ for the sealed cells. Black bow-ties are the total change in measured alkali metal relaxation, blue squares are the deduced slow portion, and red diamonds are the deduced fast portion. The x-axis error bars for the Scarecrow cell are included because the value of $[N_2]$ for this cell was estimated from it optical pumping properties and has an uncertainty of ± 0.02 bar.

In Fig. 3 the amount of change in alkali-metal relaxation as a function of N_2 pressure is plotted for the three sealed cells. As one can see, the slow component is increasing with N_2, however the fast proportion is more steady. Ref. [12] showed that the increase in alkali-metal relaxation followed a square root of neutron flux dependence. Here all of the sealed cells measured had similar amounts of captured neutron flux due to their similar ^{3}He densities, thus no corrections for neutron absorption are needed to compare the data from one cell another.

As stated earlier the valved cell data is complicated because of probable gas contamination due to the cell refilling process. This contaminant increased the alkali-metal relaxation rates in the absence of neutron flux by as much as a factor of 2 over what was expected for the measured temperature and gas fill pressures using the known Rb-Rb, Rb-^{3}He and Rb-N_2 spin destruction cross sections. Since we do not know the exact origin of the increased relaxation in the valved cell, it is impossible to surmise how it could have affected the relaxation during neutron absorption, and we show this data noting this strong caveat. As a control the valved cell was also filled without ^{3}He, i.e. only N_2 buffer gas. Without absorbed neutron flux there was no observed change in alkali-metal relaxation.

The neutron absorption is a function of $[^3\text{He}]$ where in Ref. [12] the increased alkali-metal relaxation effect was shown to scale as the square root of absorbed neutron flux. Given the range of $[^3\text{He}]$ values measured for the valved cell, these data should thus be normalized to the square root of the relative neutron absorptions, i.e. by $A_n^{1/2}$, where the values of A_n were given in Table 1. Figure 4A plots the fast and slow components and total $\Delta\Gamma$ normalized by $A_n^{1/2}$ for the valved cell data versus $[N_2]$. One may see that the slow components show a dependence on $[N_2]$. However the data from the 2.4 bar ^{3}He filling seem to contradict the hypothesis that the N_2 is mainly responsible for the increase in the magnitude of the effect given this cell had a very low $[N_2]$ but high $\Delta\Gamma$.

Perhaps the high $\Delta\Gamma$ for this cell filling is because when $[N_2]$ and/or $[^3\text{He}]$ are changed the energy captured by the gas of the cell is also changing. The absorption of neutrons by the ^{3}He creates charged hydrogen and tritium nuclei which deposit energy as they travel through the cell before either coming to rest in the gas or hitting the cell's wall, where the distance travelled

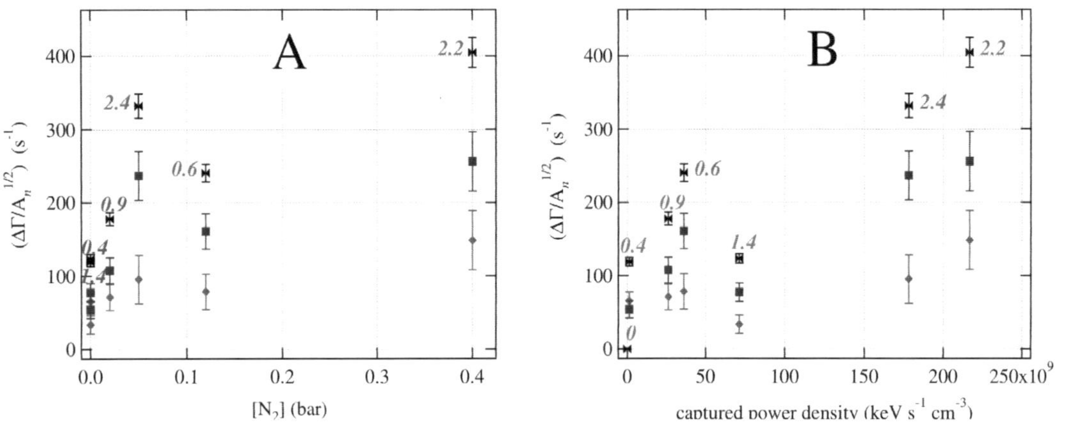

Figure 4. The measured alkali-metal relaxation versus [N_2] (graph A) and captured power density (graph B) for the valved cell with different gas fillings. In both graphs the values of $\Delta\Gamma$ have been normalized by $A_n^{1/2}$ to account for the flux dependence shown in Ref. [12]. Black bow-ties are the total change in alkali-metal relaxation, blue squares are the deduced slow portion, and red diamonds are the deduced fast portion. The numbers labelling the markers denote the ^{3}He pressure of the cell for the particular cell filling as given in Table 1.

inside the cell, and thus energy deposited in the gas, is a function of both [^{3}He] and [N_2]. This effect should also be considered in order to compare the valved cell data. Thus we also plot in Fig. 4B $\Delta\Gamma$ normalized by $A_n^{1/2}$ versus captured power density, the calculation of which is described below, for the given gas composition.

The stopping power of a gas target for charged particles is a function of the gases in the target and their density. The stopping length of the ^{1}H in one bar of ^{3}He and N_2 is 5.1 cm and 1.0 cm respectively, and for the ^{3}H the stopping length is 1.9 cm and 0.3 cm in one bar of ^{3}He and N_2 respectively. The total stopping length values L_1 for the ^{1}H and L_3 for the ^{3}H for the for cells used in this work are given in Table 1. As can be seen the size of our cells is comparable to these stopping lengths, with some values larger than the cell size and others much smaller, thus the captured power will be varying significantly over our range of cells.

The amount of captured power density, i.e. average energy per second per unit volume captured in the gas by the slowing of the charged particles created by the ^{3}He-neutron capture reaction, was therefore estimated with a simple model. The random paths of the 573 keV ^{1}H, and 191 keV ^{3}H created when a neutron is absorbed by a ^{3}He nucleus are simulated. Then given the location in the cell where they originated we calculate whether or not the particles are stopped in the gas or by the cell's wall. The particles stopped in the gas are then used to calculate the average captured power density, and the particles reaching the wall are considered to deposit their energy there. We do this because the energy deposited along the particles path is a function of particle velocity, with more energy deposited per unit length as the particle becomes slower near the end of the particle's range. Therefore since for the purposes of this discussion, where we discuss our measurements of alkali-metal relaxation measured along a line through the center of the cell, we feel this crude model is sufficient to discuss the data.

As can be seen in Fig. 4B the fast component is relatively constant over the range of captured power densities. When scaled by the square root of absorbed neutron flux the 0 N_2 cell fillings (i.e. the 0.4 bar and 1.4 bar ^{3}He points) show nearly the same increase in alkali-metal relaxation despite the different captured power densities. For the other cells it could be argued that the slow component is increasing with captured power density. Given the caveats of the valved

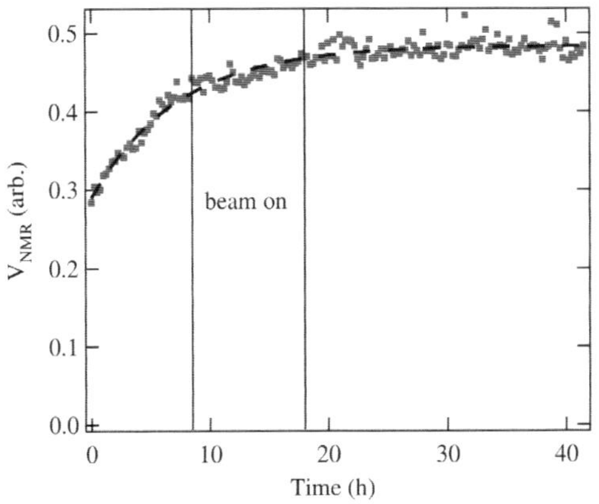

Figure 5. The measured relative ^{3}He polarization vs. time for the Roadrunner cell, a double chamber cell, via NMR free induction decay. During the time labelled "beam on" the spin filter portion of the cell was exposed to the flux of the neutron beam while continuing to polarize the cell via the spin-exchange chamber which was outside of the neutron beam. The observed ^{3}He polarization was not affected by the absorbed neutron flux in the spin filter chamber.

cell data, namely the unknown influence of the cell impurities, it is hard to draw a conclusion whether [N_2] or the captured power density is the only relevant or more important parameter from the data in Figs. 4A and 4B alone.

3. Double chambered cell

Provided the ionization effects do not cause direct mechanisms leading to nuclear spin relaxation of the ^{3}He, then a clear solution would be to use double chambered cells where the optical pumping process is isolated from the neutron absorption region. In the double chamber cell we used here, called the Roadrunner cell, the two chambers of approximately equal volumes of 30 cm^3 are connected by a small 6 mm inner diameter 5 cm long transfer tube to allow for transport of polarized ^{3}He between the chambers. As a test of this principle we installed the Roadrunner cell and polarized the spin-exchange chamber in situ while exposing the neutron spin filter chamber to the high neutron flux. The relative ^{3}He polarization was monitored using NMR free induction decay. Fig. 5 shows that the trend in the ^{3}He polarization was not modified by the neutron absorption. After this test we removed the cell and calibrated the remaining ^{3}He polarization on a different beam line were it was found to have reached over 50% during this test. Full details of this test have been described in Ref. [12].

4. Comparison to prior work and conclusions

The work conducted in the PF1B experiments, Ref. [12], measured the additional alkali-metal relaxation as a function of incident neutron flux and found a square root dependence on the absorbed flux for a given cell. This square root dependence was measured using sealed cells by varying the incident flux using calibrated neutron absorbers, keeping other parameters and experimental conditions fixed and thus was a robust measurement. Most of the cells in this work had similar levels of neutron absorption and also had similar values of [N_2], ranging from 0.07 bar to 0.12 bar, except for one cell which had 0.5 bar N_2. The cell with 0.5 bar N_2 clearly showed a much higher level of neutron induced alkali-metal relaxation. This indicates at least two dependent parameters to consider when comparing the two experiments.

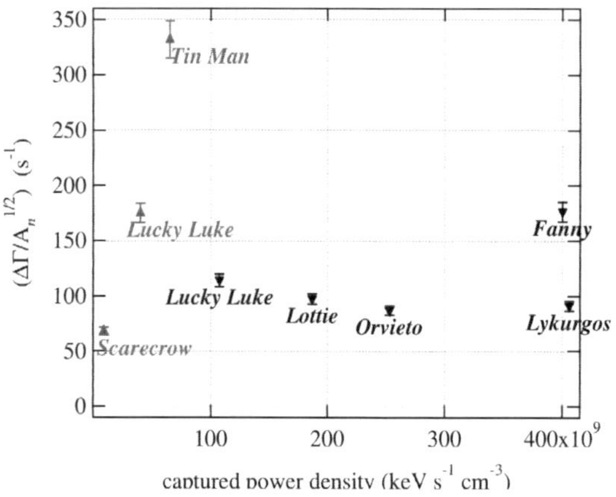

Figure 6. The measured change in alkali-metal relaxation verses captured power density for the cells from the PB1B experiment [12] (black downward triangles) and the sealed cells from this work (red upward triangle). All of the values of $\Delta\Gamma$ have been normalized to the square root of the normalized absorbed neutron flux, where the normal neutron flux (i.e. flux=1) was taken as the full flux of the tomography beam line used in this work.

The data for the Lucky Luke cell from this experiment can be compared with its data from the PF1B experiment because it was measured in both beams. When the values for the Lucky Luke cell from the PF1B data are fit to the square root of absorbed neutron flux, the fit values would predict a measured increase in alkali-metal relaxation of 67 s^{-1} for the measured neutron flux of the tomography beam line and ^{3}He density of the Lucky Luke cell. We measured 100 s^{-1} $\pm$ 5% and feel this is reasonable agreement given the different conditions of the two experiments. We can surmise the limited knowledge of the spectrum of the tomography beam used here, or some other factor such as perhaps sampling a different location in the cell with our probe beam could lead to such a difference.

In Fig. 6 we plot the data from the the sealed cells in this work and the data from PF1B for $\Delta\Gamma/A_n^{1/2}$ versus captured power density. To account for the differences in absorbed neutron flux from the two beams we have used the measured neutron flux of the tomography beam in this work as a normalization constant. The relative neutron absorptions of the cells used in the PF1B experiment, for the energy spectrum of that beam, were given in Ref. [12], then given the flux on PF1B was 4.7/2.4 or 1.96 times that of the tomography beam, one can normalize the PF1B data to the square root of the absorbed neutron flux relative to the flux of the tomography beam. In this plot we can see the data from the PF1B experiment did not show a dependence on the captured power density, the only indication was of a dependence of the effect on $[N_2]$. The magnitude of the neutron induced relaxation observed here, is slightly higher, after scaling to the square root of absorbed neutron flux as stated above, but within reason given differences in the two experiments.

In conclusion, it is still not possible to determine the definitive origin of the mechanism involved in the unexpectedly high alkali-metal relaxation observed in high flux neutron beams. The measurement of the square root dependence was not repeated during this study but is assumed from the prior work and used to compare the data obtained in this work. We have confirmed that the neutron induced alkali-metal relaxation is increasing with $[N_2]$ but it is unclear if this is due to a relaxation mechanism involving the nitrogen directly, or because of the effect of nitrogen density on the charged particle stopping power of the cell; the data from the

PF1B work and the valved cell data in this work may lead one to different conclusions. We have confirmed the presence of two time scales for neutron induced alkali-metal relaxation. The fast component happens on a time scale of less than 1 s and seems relatively constant with respect to [N_2] or captured power density. The slow component is increasing strongly with respect to captured power density or [N_2].

It is hard to imagine a way to eliminate this effect given the fast component alone such that efficient optical pumping could be done under conditions of absorption of extremely high neutron flux. At moderate flux levels the effect could likely be overcome with higher power optical pumping sources. Whereas at the highest neutron flux we believe the practical solution will be to perform the polarization of the ^{3}He outside of the neutron beam. With SEOP this can be accomplished in an in situ polarizer using a double chambered cell following the success using this technique for polarized electron scattering targets [18].

5. Acknowledgments

We would like to thank the ILL direction for making this beam line available for this non-standard experiment. We would also like to thank the glass workshop in the ZAT part of the Institut für Festkörperforschung, Forschungszentrum Jülich GmbH for making the valved cell used in this work. We also thank the ILL ^{3}He team for technical support during this experiment. T.R Gentile, G.L. Jones, and T.G. Walker received support from the US Department of Energy, Office of Basic Energy Sciences, Division of Materials Sciences and Engineering under Grant No. DE-FG02-03ER46093.

References

[1] Andersen K H, Jullien D, Petoukhov A K, Mouveau P, Bordenave F, Thomas F and Babcock E 2009 *Physica B-Condensed Matter* **404** 2652 – 2654
[2] Chen W C, Erwin R, McIver-III J W, Watson S, Fu C B, Gentile T R, Borchers J A, Lynn J W and Jones G L 2009 *Physica B-Condensed matter* **404** 2663–2666
[3] Nacher P J and Leduc M 1985 *Jounal de Physique* **46** 2057–2073
[4] Chann B, Babcock E, Anderson L W, Walker T G, Chen W C, Smith T B, Thompson A K and Gentile T R 2003 *J. of Appl. Phys.* **94** 6908–6914
[5] Babcock E, Chann B, Nelson I A and Walker T G 2005 *App. Optics* **44** 3098–3104
[6] Babcock E, Nelson I A, Kadlecek S, Driehuys B, Anderson L W, Hersman F W and Walker T G 2003 *Phys. Rev. Lett.* **91** 123003
[7] Chen W C, Gentile T R, Walker T G and Babcock E 2007 *Phys. Rev. A* **75** 013416
[8] Chen W C, Armstrong G, Chen Y, Collett B, Erwin R, Gentile T R, Jones G L, Lynn J W, McKenney S and Steinberg J E 2007 *Physica B Condensed Matter* **397** 168–171
[9] Babcock E, Mattauch S and Ioffe A 2011 *Nuclear Instruments and Methods A* **625** 43 – 46
[10] Chen W C, Gentile T R, Fu C B, Watson S, Jones G L, McIver J W and Rich D R 2011 *submitted to IOP journal of physics conference series, this issue*
[11] Abele H, Dubbers D, Hase H, Klein M, Knopfler A, Kreuz M, Lauer T, Markisch B, Mund D, Nesvizhevsky V, Petoukhov A, Schmidt C, Schumann M and Soldner T 2006 *Nuclear Instruments and Methods A* **562** 407–417
[12] Babcock E, Boag S, Becker M, Chen W C, Chupp T E, Gentile T R, Jones G L, Petukhov A K, Soldner T and Walker T G 2009 *Phys. Rev. A* **80** 033414
[13] Bonin K D, Walker T G and Happer W 1988 *Phys. Rev. A* **37** 3270–3282
[14] Coulter K P, McDonald A B, Cates G D and Happer W 1989 *Nuclear Instruments and Methods A* **276** 29–34
[15] Babcock E, Boag S, Andersen K H, Becker M, Beechamb C, Bordenave F, Chastagnier J, Chen W C, Chung R, Chupp T E, Elmore S, Fouilloux P, Gentile T R, Jullien D, Lelievre-Berna E, Mouveau P, Petoukhov A, Revert M and Soldner T 2009 *Physica B-Condensed Matter* **404** 2655–2658
[16] Franzen W 1959 *Physical Review* **115** 850–859
[17] Wagshul M E and Chupp T E 1994 *Phys. Rev. A* **49** 3854–3869
[18] Singh J, Dolph P, Mooney K, Nelyubin V, Tobias A, Kelleher A, Averettt T and Cates G 2009 *SPIN PHYSICS Book Series: AIP Conference Proceedings* **1149** 823–828

Journal of Physics: Conference Series **294** (2011) 012012 doi:10.1088/1742-6596/294/1/012012

^{3}He spin filters for spherical neutron polarimetry at the hot neutrons single crystal diffractometer POLI-HEiDi

V Hutanu[1*], M Meven[2], S Masalovich[2], G Heger[1], and G Roth[1]
[1]Institute for Crystallography, RWTH Aachen University, 52056 Aachen, Germany
[2]FRM II, Technische Univerisität München, 85747 Garching, Germany

E-mail: vladimir.hutanu@frm2.tum.de

Abstract. 3D vector polarisation analysis called also SNP (Spherical Neutron Polarimetry) is a powerful method for the detailed investigation of complex magnetic structures. The precise control of the incoming and scattered neutron polarisations is essential for this technique. Here we show an instrumental setup, that was recently implemented on the new single crystal diffractometer POLI-HEiDi at the FRM II for performing SNP experiments using two ^{3}He spin filters for the production and for the analysis of the neutron polarisation. The design and optimisation procedure for the used spin filter cells are presented. Methods for in-situ measurements of the incoming polarisation as well as the particularities of the using two spin filters and corrections for the time dependent relaxation are discussed. Statistical precision of 1% has been achieved for the measurements of the polarisation matrix under the real experimental conditions using described cells and applying proposed correction method for the data.

1. Introduction

Gaseous ^{3}He neutron Spin Filter Cells (SFC) are simple transmissions devices, with predictable characteristics, that do not require any special optical adjustments on a neutron beam. Since polarized nuclei of ^{3}He possess a very high spin-dependent neutron absorption efficiency over a wide range of energies, the ^{3}He SFC can be used as a broadband neutron polariser or analyser, with the possibility to optimise its efficiency for nearly all neutron wavelengths. The use of ^{3}He SFC allows for the decoupling the processes of neutron reflection and polarisation, unlike the case of the Heussler crystal polariser. The straight line transmission of the SFC allows the use of a focussed beam and gives a homogenous polarisation over the entire beam cross section. Up to now these advantages have been mostly used to produce SFC analysers especially for small angle neutron scattering or reflectometry [1, 2, 3]. However in the combination with hot neutrons SFC is a particularly efficient polarising device. According to the calculations [4] the combination of the focused Cu [220] crystal monochromator with ^{3}He spin filter with 70% ^{3}He polarisation will increase the efficiency of the instrument in the production of the hot polarised neutron beam by a factor of 1.7 compared to the Heussler crystal monochromator. Exactly this design has been employed in the new polarised diffractometer POLI-HEiDi. The advantage of this technique is a better resolution and a high flux of the polarised short wavelength neutrons on the sample. The drawback is a degradation of the incident neutron beam polarisation and analysing efficiency of the analyser with time. Really, when using one SFC as an analyser the measured resulting polarisation is proportional to the time dependent analysing efficiency of the cell. Therefore, precise knoweledge of cells parameters allows for straightforward corrections of the data for this time dependence. However, when using two SFC as polariser and analyser in the same time, the measured resulting polarisation is proportional to a product of two independent time functions, so that the correction is only possible if polarisations of the ^{3}He gas in the polariser and analyser cells are measured independently. A way how to perform this is presented in the Sect. 4. In the Sect. 3 we describe the optimisation of cell's parameters. Important to note that using the optimal cells parameters in order to maximise the figure of merit Q and not the polarising efficiency P_n lead to better results regarding the precision of experimental measurements. In the present report we demonstrate that by using the optimal SFC, the precise control of the incoming polarisation and accurately performed corrections for the time dependent depolarisation, the measurements of the polarisation matrices with the required precision are implemented on the new diffractometer POLI-HEiDi.

2. POLI-HEiDi diffractometer

The polarisation of the neutron beam can be described in terms of the classical vector and can be tuned in the frame of the instrument coordinate system. 3D vector polarisation analysis known also as Spherical Neutron Polarimetry (SNP) is a technique where all components of the scattered polarisation vector are measured in turn for three different directions of the incoming polarisation vector. Determining the relationship between the directions of incident and scattered polarisations gives access to the 16 independent correlation functions involved in the most general nuclear and magnetic scattering process. Generally, this allows for the determination of the direction of the magnetic interaction vectors of magnetic structures. For those structures, in which nuclear and magnetic reflections coincide in reciprocal space, SNP leads to the determination of the amplitude of the magnetic interaction vectors and hence to the magnetisation distribution [5]. SNP using third generation zero-field polarimeter CRYOPAD [6] have been recently implemented and successfully tested on the new single crystal diffractometer POLI-HEiDi at the FRM II [7,8]. The unique feature of the POLI-HEiDi compared to other instruments performing SNP is that both neutron polarisation production and analysis are done with ^{3}He spin filters. The parameters of the used spin filters are optimised in order to achieve the best performance in measuring polarisation at a given wavelength. Special DECPOL-like magnetostatic cavities

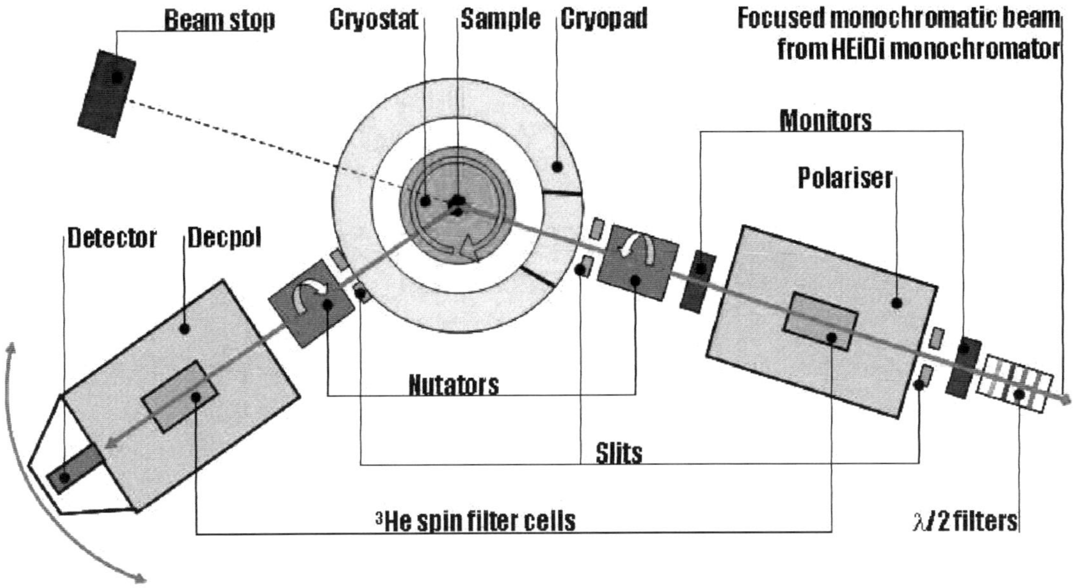

Figure 1. (colour on-line). Instrument layout of the SNP - setup on POLI-HEiDi.

have been constructed and built for the housing the cells during the experiment. The layout of the instrument is shown in the Fig. 1. Non-polarised focused monochromatic neutron beam (different wavelengths are available) from the HEiDi monochromator is coming from the right in the picture (see [9] for more details).

3. Spin filter cells

It has been shown in [10] that the statistical accuracy in the determination of the scattered beam polarisation is directly proportional to the square root of the quality factor Q, of the spin filter. The Q value is denoted in terms of neutron polarisation P_n and transmission T_n as:

$$Q = P_n^2 T_n \, . \tag{1}$$

The maximisation of Q leads to the optimisation of the spin filter parameters. Usually spin filter cells are characterised by their opacity O denoted as:

$$O = 0.0732 \; p[\text{bar}] \; l[\text{cm}] \; \lambda[\text{Å}] = 0.0732 \; O', \tag{2}$$

where p is the pressure of the ^{3}He gas in the filter at 20°C, l is the length of the neutron path through the gas, and λ is the neutron wavelength. According to [11] the optimum opacity O'_m at given P_{He} is:

$$O'_m = \frac{1}{P_{He}} \operatorname{arctan}h\left(-\frac{1}{2P_{He}} + \sqrt{\frac{1}{4P_{He}^2} + 2}\right) \frac{\text{bar cm Å}}{0.0732} \, . \tag{3}$$

Taking $P_{He} = 0.7$ as a realistic mean value that is routinely delivered by the HELIOS filling station at the FRM II [12], one obtains a simple expression for the optimal opacity of the used spin filters:

$$p[\text{bar}] \; l[\text{cm}] \; \lambda[\text{Å}] \approx 26. \tag{4}$$

For a certain wavelength, the only parameters to optimise are the length of the cell l and the gas pressure p. In the following, we will define the meaningful limits for both. It is worth noting, that we considered so far optimal conditions for the case P_{He} = const. However, after a cell had been detached from a filling station, the P_{He} in the cell on the neutron instrument decreases

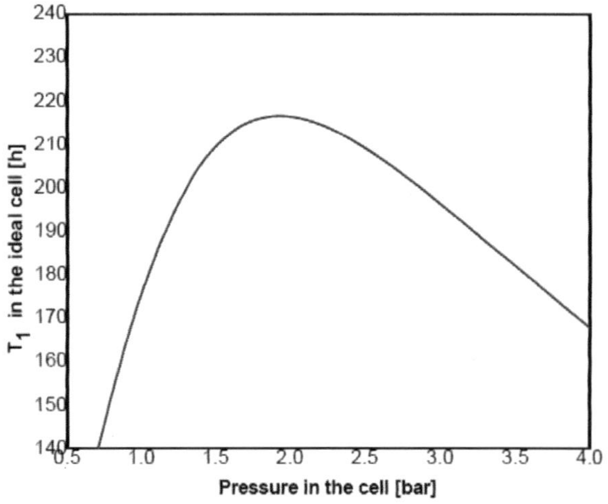

Figure 2. Pressure dependence of T_1 in a surface relaxation free cell situated in a magnetic field with transversal gradient of 5 x 10^{-4} cm^{-1}.

exponentially with the time constant T_1 (relaxation time):

$$P_{He}(t) = P_{He}(0)\exp(-t/T_1)\ . \qquad (5)$$

Therefore, beside geometrical parameters of the cell, T_1 is another essential parameter for the spin filter. Three major relaxation mechanisms are generally accepted: 1) surface relaxation on a filter walls, 2) relaxation due to a magnetic field gradient $\nabla|\vec{B}_\perp|/B_0$, in a magnetostatic cavity, 3) relaxation due to the dipole-dipole interaction inside the polarised gas. The last two mechanisms are pressure dependent. Considering an ideal surface relaxation free cell, T_1 can be written as:

$$\frac{1}{T_1[h]} = 1.7\times10^4\cdot(\nabla|\vec{B}_\perp|/B_0)^2\frac{1}{p} + \frac{p}{750}\ . \qquad (6)$$

By taking for the field gradient a realistic value of 5 x 10^{-4} cm^{-1} one obtains the pressure dependence of T_1 shown in Fig. 2. It is easy to see that T_1 reaches the maximum at the pressure close to 2 bar. We assume now that the same cell will be used in experiments with different neutron wavelengths and the optimization will be done by tuning the pressure. If we accept that the reduction of T_1 due to pressure tuning should not overcome 10% of T_1^{max}, then we come to the conclusion that the optimal pressure region expands from 1.3 to 3 bar for a gradient of 5 x 10^{-4} cm^{-1}.

Figure 3. (colour on-line). 130 mm long and 60 mm in diameter SFC used at POLI-HEiDi. The cells are made of high purity quartz glass and coated inside with metallic Cs. Small Cs droplets can be observed in the lower part and in the appendix of the cell on the left. The dark spots inside the cell on the right are Cs sub-oxides, which are liquid at room temperature.

Now the length of the cells requested for the operation with neutrons with wavelengths from 0.55 Å to 1.17 Å can be easily determined from Eq. 4. The following practical limitations have to be considered as well: a) the geometry of the polariser and analyser cells should be the same (exchangeability), b) the space available at the instrument for the polariser magnetostatic cavity is limited to max. 50 cm - this means that cells should be as short as possible to achieve a high magnetic field homogeneity . Taking into account all mentioned criteria, cells with length l = 13 cm have been produced for the POLI-HEiDi. The cell diameter should be large enough to cover whole incoming beam reaching the sample position as well as the complete scattered beam reaching the detector. Considering the beam spot at the sample position of about 30 mm in diameter and the convergent incoming beam, cells of 60 mm in diameter have been designed. Typical cells used on POLI-HEiDi are shown in Fig. 3. Fig. 4 presents the calculated quality factor of our cells for two different filling pressures at P_{He} = 0.7. Maximising of the Q for almost all available on HEiDi monochromator wavelengths is possible using the same cell by tuning the pressure.

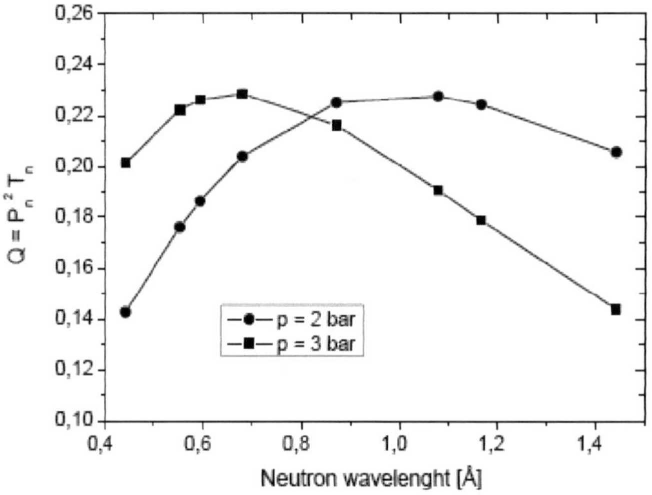

Figure 4. Quality factor of the used cells calculated for two different pressures and P_{He} = 0.7. The filled data points correspond to available discrete wavelengths, the connecting lines are the eye-guides showing the wavelength dependent behaviour.

The cell windows are made from the same type of fused silica material as the cell body and are 3-4 mm thick in order to withstand the pressure up to 3 bar. They are welded to the cells body. The neuron transmission of empty cells T_0, measured for the used wavelengths is about 0.86-0.88 depending on the window thickness. Relaxation times T_1 measured in different cells after the preparation varies between 100 h and 200 h, when T_1 up to 260 h have been measured in some of the cells. T_1 of any particular cell can also vary with time depending on the number of cell fillings, slow Cs oxidation, magnetic history of the cell, etc.

4. Polarisation measurements and time-dependent corrections

In the preceding sections we discussed following advantages of the ^{3}He SFC: high efficiency for hot neutrons, easy adjustment for the different wavelengths, flexibility regarding beam geometry, etc. The importance of the T_1 has been also mentioned. Time dependency of the polarising (analysing) efficiency of the filters represents a main drawback of the technique requiring additional corrections, which lead to some decrease of the statistical accuracy of the measurement. Since the precise knowledge of the incoming and scattered beam polarisation is essential for the SNP, in the following we will describe briefly the correction procedure applied to the data measured on POLI-HEiDi. We note that the accurate error analysis of the correction procedure is not the subject of the present report and it will be treated separately.

The correction is possible in principle due to the fact that the filter's performance follows the known theoretical predictions. In order to apply them, two conditions should be satisfied for each measurement. First, the precise knowledge of the filter parameters such as opacity and T_1 is required both for polariser and analyser. Second, a time mark should be assigned to each measured experimental value.

We denote with n^+ and n^- the counting rates for the neutrons with the parallel and antiparallel direction of the polarisation vector with respect to the magnetic field in the analyser, respectively. The experimentally measured asymmetry A

$$A = \frac{n^+ - n^-}{n^+ + n^-} \tag{7}$$

has the meaning of the product of scattered neutron beam polarisation P' and analysing efficiency of the analyser A_n:

$$A = P' A_n . \tag{8}$$

Measuring n^+ at the time t_1 and n^- at the time t_2 ($(t_2-t_1) << T_1$), we define the scattered beam polarisa-tion at the mean time $t_m = (t_1+t_2)/2$, $P'(t_m)$ as:

$$P'(t_m) = \frac{n^+(t_1) - n^-(t_2)}{n^+(t_1) + n^-(t_2)} \cdot \frac{1}{A_n(t_m)} \tag{9}$$

In SNP, the usual experimental strategy is to measure the scattered beam polarisation P' when the incident polarisation P is set alternatively along x, y or z axis. This determines the polarisation matrix. The matrix element P'_{ij} gives the ith component of the scattered polarisation vector when the incident polarisation is in the jth direction. Using the time marking and the same time scale for the polariser and analyser, the precise measurement of P'_{ij} is getting possible. The incident polarisation $P(t_m)$ at the moment t_m can be precisely calculated using transmission measurements from two neutron monitors, placed before and after the polariser (see Fig. 1). Transmission of the polariser T^p is a simple normalised ratio between the two monitors:

$$T^p(t) = M_2(t) / M_1(t). \tag{10}$$

Fig. 5 shows an example of the time-dependent measured transmission in one particular experiment. Transmission of the spin filter calculated for a non-polarised beam depends on a ^{3}He gas polarisation in the cell as:

$$T^p(t) = T_0^P e^{-O^P} \cosh(O^P P_{He}^P(t)) \tag{11}$$

where T_0^p is the transmission of the empty polariser cell, O^p is the opacity of the polariser cell defined by Eq. 2, $P_{He}^p(t)$ is the time dependent polarisation of ^{3}He gas in the polariser cell defined by Eq. 5. Fitting the experimental transmission data (Fig. 5) using Eq. 11 results in precise values for the ^{3}He initial polarisation and T_1 of the polariser cell. The refined values for the presented experiment are shown in the inset of Fig. 5: $P_{He}^p(0) = 0.7363(3)$, $T_1^p = 93.77(22)$ h. In Fig. 6 the calculated time-dependent polarisation of the incoming beam is shown for the data plotted on Fig. 5. The measurement of the scattered polarisation matrix element P'_{ij} according to Eq. 9 requires a precise knowledge of the time-dependent analysing efficiency $A_n(t)$:

$$A_n(t) = \tanh\left(O^a P_{He}^a(0) \exp(-t/T_1^a)\right) , \tag{12}$$

where O^a is the opacity of the analyser cell, $P_{He}{}^a(0)$ is the polarisation of the ^{3}He gas at the time moment zero (initial gas polarisation), $T_1{}^a$ is the relaxation time constant of the analyser.

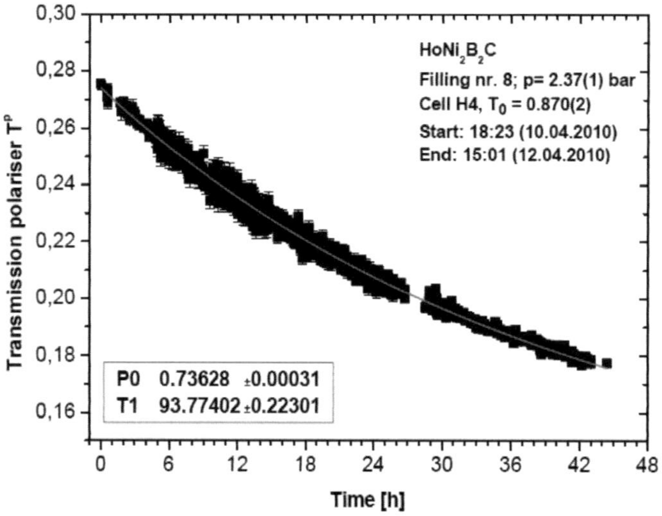

Figure 5. Time-dependent transmission of the polariser cell measured using two transmission monitors situated before and after the polariser. The result of the least squares fit of ^{3}He initial polarisation and T_1 according to Eqs. 5 and 11, are given in the inset.

Using known O^a the problem reduces to the determination of the initial polarisation and $T_1{}^a$ in the analyser. One can use different ways to do this. First, by analogy with the procedure described for the polariser, one can measure transmission of the analyser cell with the incident non-polarised beam. This has been done in our first SNP experiment with two spin filters using MuPAD [12], where the polariser cell had to be removed temporally from the incoming beam to perform the measurement only with the analyser cell. This is not technically foreseen in the actual POLI-HEiDi design, therefore it could be only done for two time points: before the measurement when the polariser cell is not yet installed into the incoming beam and after the measurement, when the polariser cell is removed from the beam for the replacement with the "fresh" polarised gas. Knowing transmissions over a definite time interval and assuming that depolarisation follows the form given by Eq. 5, one can calculate $P_{He}{}^a(0)$ and $T_1{}^a$ using an expression analogous to Eq. 11. However, as shown in practice, the values derived only from two measurements have a rather high uncertainty.

Another method that allows time-dependent corrections to measured data is based on the transmission measurement of the partially polarised beam according to expression derived in Ref. [13]:

$$T_a(t) = T_0{}^a \exp(-O^a) \cosh(O^a P_{He}{}^a(t)) \, [1 + P_n(t) \tanh(O^a P_{He}{}^a(t))] \,, \tag{13}$$

where $P_n(t)$ is the time-dependent polarisation of the incident beam (see Fig. 6). Although measurements in arbitrary number of points is possible and the subsequent least square fit will improve the accuracy of the refined parameters, this is nevertheless an indirect method, which also includes the uncertainty in the calculation of $P_n(t)$. Moreover, the transmission measurements should be done in the direct beam or at least on a pure nuclear Bragg reflection in order to be sure that the polarisation of the incoming beam is not changed due to the interaction with the sample.

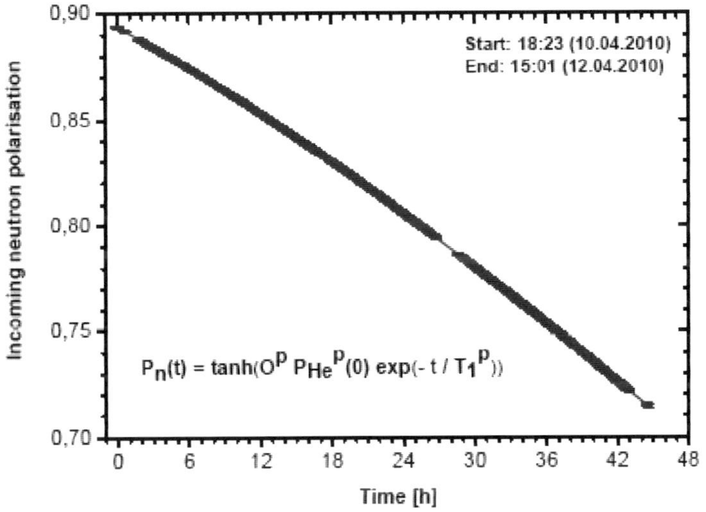

Figure 6. Time-dependent polarisation of the incoming beam *P(t)* derived from the transmission data in Fig. 5. The thick line in the graph includes the error bars.

Time dependent measurements of non-magnetic Bragg reflection permits, however, a direct measurement of $A_n(t)$. In this particular case $A_n(t)$ can be easily found by normalising the measured asymmetry (Eqs. 7 and 8) to the incoming polarisation measured earlier. Taking the predefined O^a to be a fixed parameter and fitting $A_n(t)$ to measured data with Eq. 12, one obtains the required parameters of the analyser cell: $P_{He}{}^a(0)$ and T_1^a. Fig. 7 shows an example of the time-dependent analyser efficiency calculated for the set of experimental data presented in Fig. 5 and 6 measured with a nuclear Bragg peak. Due to high counting rates in nuclear Bragg peaks, the total time spent for additional measurements necessary for the correction parameter determination is less than 5% of the total measurement time.

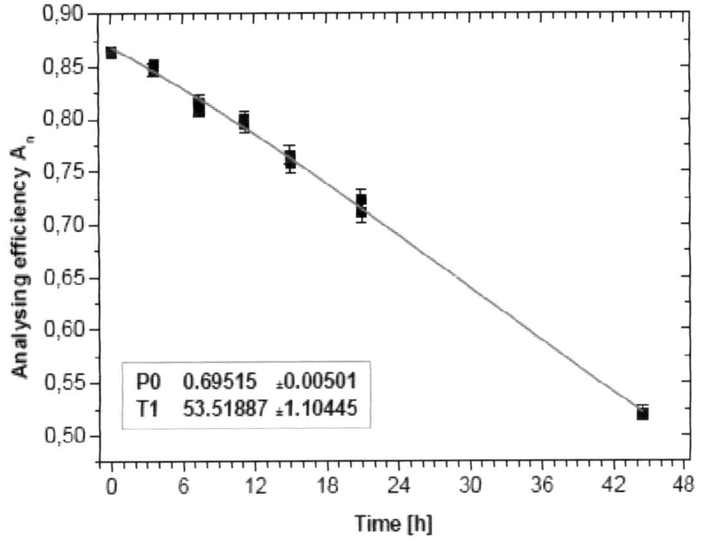

Figure 7. Measured time dependent analysing efficiency of an analyser cell. The result of the least squares fit gives the ^{3}He initial polarisation and T_1 according to Eqs. 5 and 12. The reason for the unusually low T_1 observed in the DECPOL [8] in this measurement is a wrong current setting in the coils.

5. Results and conclusion

Measured and corrected polarisation matrices for two Bragg reflections [200] and [005] are presented in Table 1. The measuremets have been done on a $HoNi_2B_2C$ superconducting sample at 4.2K at the POLI-HEiDi diffractometer using the described setup and neutrons with the wavelength of 0.87 Å. The Bragg peak [200] from this sample is a purely nuclear one and thus a singular diagonal matrix has been measured with a high precision. Conversely, the [005] peak is a purely magnetic reflection, and rotation through 180° for the P_{xx} component is seen. Other diagonal components are totally depolarised that suggests the presence of the equally populated 180° magnetic domains in the structure. All diagonal and off-diagonal components could be determined with the precision of 1% using the described correction procedure.

Table 1. Polarisation matrices for two different Bragg peaks from a $HoNi_2B_2C$ sample "as measured" and corrected for ^{3}He time relaxation.

Bragg peak	P_{in}. / P_{out}.	Asymmetry as measured			Corrected polarisation matrix		
		x	y	z	x	y	z
2 0 0 nuclear	x	0,730(3)	0,003(5)	0,009(5)	0,99(1)	-0,00(1)	0,01(1)
	y	0,030(5)	0,714(3)	-0,006(5)	0,04(1)	0,98(1)	-0,01(1)
	z	0.001(5)	0,005(3)	0,715(3)	-0,00(1)	0,00(1)	0,99(1)
0 0 5 magnetic	x	-0,652(12)	0,004(11)	-0,032(11)	-1,00(3)	0,01(1)	-0,06(1)
	y	0,004(11)	0,004(11)	-0,094(11)	0,01(1)	0,01(1)	-0,10(1)
	z	-0,020(11)	-0,095(11)	0,009(11)	-0,05(1)	-0,11(1)	-0,00(1)

The SFC with the optimised parameters are employed to produce and analyse neutron polarisation at the new polarised diffractometer POLI-HEiDi (RWTH Aachen) at FRM II. The use of SFC assures a high resolution and the optimal efficiency for hot neutrons. Spherical neutron polarimetry using the third generation zero-field polarimeter CRYOPAD for the investigation of complex magnetic structures is implemented on POLI-HEiDi. The correction procedure for the time-dependent polariser (analyser) efficiency has been developed and successfully applied.

Acknowledgements:
The work has been supported by the German Federal Ministry for Education and Science (BMBF) through the projects 03HE7AAC and 03HE6AA3.

References:

[1] Lelievre-Berna E et al. 2005 *Physica B: Condensed Matter* **356** 141.
[2] Keiderling U, Wiedenmann A, Rupp A, Klenke J and Heil W 2008 *Meas. Sci. Technol.* **19** 034009

[3] Chen W C, Erwin R, McIver J W, Watson S, Fu S B, Gentile T R, Borchers J A, Lynn J W, and Jones G L, 2009 *Physica B: Condensed Matter* **404** 2663
[4] Lelievre-Berna E and Tasset F, 1999 *Physica* B*: Condensed Matter* **267** 21
[5] Brown P J, 2005 Spherical Neutron Polarimetry, in *Neutron Scattering From Magnetic Materials*, ed. Chatterji T
[6] Lelievre-Berna E, Anderson I and Guérard B , Eds., 2002 *Proceedings of SPIE* **4785** 112
[7] Hutanu V, Meven M and Heger G, 2007 *Physica B: Condensed Matter* **397** 135
[8] Hutanu V, Meven M, Lelievre-Berna E and Heger G, 2009 *Physica B* **404** 2633
[9] http://www.frm2.tum.de/wissenschaftliche-nutzung/diffraktion/poli-heidi/index.html.
[10] Tasset F and Ressouche E, 1995 *Nucl. Instr. and Meth..* **A 359** 537
[11] Surkau R et al. 1995 *Nucl. Instr. and Meth* A **384** 444
[12] Hutanu V, Masalovich S, Meven M, Lykhvar O, Borchert G and Heger G, 2007 *Neutron News* **18** 14
[13] Hutanu V, Janoschek M, Meven M, Böni P and Heger G 2009 *Nucl. Inst. and Meth.* A **612** 155
[14] Masalovich S, *Nucl. Inst. and Meth.* A **581** 791

Journal of Physics: Conference Series **294** (2011) 012013 doi:10.1088/1742-6596/294/1/012013

Wide-angle polarization analysis with ^{3}He for neutron scattering instrumentation at the JCNS

A Ioffe*, E Babcock, V Pipich, A Radulescu
Jülich Centre for Neutron Science at FRM II, Forschungszentrum Jülich GmbH, 85747 Garching, Germany

E-mail: a.ioffe@fz-juelich.de

Abstract. Polarization analysis is an important technique for polarized neutron scattering as it allows one to obtain the full information about the vector magnetization in the sample that is critically important for detailed understanding of physical properties of molecular magnets, new superconductors, spin electronic and magnetic nanostructures, as well as the self-organization of magnetic nanostructures. In the simplified 1-dimensional version polarization analysis allows for the separation of coherent and incoherent scattering, making it a potentially important technique for studies of non-deuterated biological objects that themselves produce unavoidable background. We compare some of the major considerations between two different methods for the polarization analysis – supermirror based analyzers and polarized ^{3}He neutron spin filters and point out when the latter is beneficial from the point of view of our neutron experiments and instrumentation. We will also discuss some specific requirements to such neutron spin filters and summarize the classes of instrumentation where they will be applied at the JCNS. Finally we will describe a successful application for small-angle neutron scattering from a biological sample.

1. Introduction

Detailed understanding of physical properties of new substances such as molecular magnets, new superconductors, spin electronic and magnetic nanostructures, as well as mechanisms involved in their formation, e.g. the self-organization of magnetic nanostructures, requires the full information about the vector magnetization in the sample, i.e. both about the magnitude and the spatial direction of the magnetic field.

Generally speaking, the ultimate task of polarized neutron scattering experiments is to determine the depolarization tensor D that defines the polarization of the scattered beam P_{ij} after the scattering

$$P_{ij} = DP_i \tag{1}$$

where P_i ($i = x, y, z$) are the values of the three orthogonal components of the neutron spin in the incident beam and P_{ij} ($i, j = x, y, z$) are the values of the polarization measured for i-th component of the neutron spin in the incident beam and for j-th component of the neutron spin in the scattered beam. The polarization tensor is given by the matrix [1, 2]

$$D = \begin{pmatrix} P_{xx} & P_{xy} & P_{xz} \\ P_{yx} & P_{yy} & P_{yz} \\ P_{zx} & P_{zy} & P_{zz} \end{pmatrix} \tag{2}$$

To perform 1-dimensional polarization analysis one determines only the polarization P_{zz} (z – the quantization axis) by measurements of spin-up and spin-down scattering intensities. However, even in this simplified form polarized neutron techniques are very useful. For example, 1-dimensional PA allows for the separation of (weak) coherent and incoherent scattering from the sample (see Sect. 4).

Another technique is the so-called full polarization analysis (or XYZ polarization analysis) when one determines the diagonal elements of the matrix D (Eq. 2). In this case 1-dimensional polarization analysis is separately performed for all 3 components (x, y and z, i.e. for two spin directions normal to the quantization axis and one parallel to it, respectively) of the incident beam polarization and the 3 values P_{xx}, P_{yy} and P_{zz} are determined. With this method the magnetic scattering cross-sections can be separated from the nuclear scattering in the scattering from the sample in addition to the abovementioned separation of the incoherent and coherent scattering cross sections.

Finally, using the vector polarization analysis (or as it is also called spherical polarization analysis) one can also determine the off-diagonal elements of the matrix D (Eq. 2) and establish the orientation of the magnetic field vector [1, 2]. Moreover, a spread in the orientation of the magnetic field vector leads to a spread in the orientation of the outgoing neutron spin vector causing the depolarization of the incident neutron beam. Thus, the vector analysis of the neutron polarization together with neutron depolarization measurements may provide full information, about the magnetic field distribution (both amplitude and direction) in mesoscopic/nanoscopic magnetic structures.

All these applications require the polarization analysis of the neutron beam scattered by the sample. Dictated by the limited brilliance of neutron sources, a large number of new neutron scattering instruments, both recently built and in the design phase, are equipped with large area position-sensitive detectors to compensate for the low brilliance and/or low scattering intensity by measuring as much of the scattering pattern at once as possible (see [3-6] for some examples). This means that the polarization analysis of the scattered beam should also be performed over the wide scattering angle spanned by such detectors in order to use the instrument effectively.

Moreover, as the new generation of neutron sources (SNS, J-PARC and ESS) are spallation sources, which produce relatively short high intensity bursts of neutrons of a broad energy spectrum, the neutron energy can then be determined via the different source to sample flight times for neutrons of different energies. Thus the neutron instrumentation designed for these sources aims to fully exploit the time structure of their neutron beams and must be able to provide efficient polarization analysis over the full neutron energy spectrum produced in each neutron pulse.

Wide area neutron detectors are used differently depending on the type of measurement performed. The value of the maximum scattering angles Ω_h and Ω_w (here indices h and w denote the height and the width) to be covered by the analyzer differs significantly for different neutron scattering instruments (see Table 1). They range from about 3° for high Q-resolution small angle neutron scattering (SANS) diffractometers to about 190° for diffractometers and time-of-flight spectrometers. As an example, values of solid angles (Ω_h x Ω_w) for certain neutron scattering instruments [7] at the JCNS are presented in Table 1.

Besides the rather varied requirements of the solid angles for different instruments, there are also some specific requirements to the required quality of neutron spin analyzers that they should not degrade the general neutron performance of the instrument, in particular they should not spoil the instrument resolution. This leads one to prefer polarized ^{3}He neutron spin filters (^{3}He NSF) over supermirror analyzers in many instances because ^{3}He NSFs are normally considered not to alter the neutron beam divergence.

Table 1. Parameters of the sample-detector parts of different types of neutron scattering instruments operated by JCNS at FRM II [7].

	Detector size (height x width)	Sample-detector distance	Solid angle (Ω_h x Ω_w)
TOF spectrometers (TOPAS)	2 m x 9 m	3 m	45° x 150°
Diffractometers (DNS)	1 m x 5 m	1.5 m	33° x 190°
Reflectometers (MARIA)	0.4 m x 0.4 m	2 m	11° x 11°
Low Q-resolution SANS (KWS2)	1 m x 1 m	1 m	45° x45°
High Q-resolution SANS (KWS1)	1 m x 1 m	20 m	3° x 3°

However, the neutron optical quality of the ^{3}He cell material must not be overlooked: in cases where the scattering angle is very small or the detector spatial resolution is very high the scattering from the surfaces of the ^{3}He NSF also needs to be considered (this will be discussed in Sect. 3).

In this article we will consider different areas where wide-angle polarization using ^{3}He spin filters is beneficial from our point of view, discuss some aspects of their implementation and give some practical examples. For purposes of the proceedings of this workshop we will give some general conceptual arguments aimed for a broad audience without intending to give a comprehensive coverage of the topic of wide-angle polarization analysis. Further explanations and more detailed discussions can be found in the given references and the references therein. Aiming to discuss polarization analysis for large area position-sensitive detectors, we will not discuss some common applications of polarization analysis (e.g. for three-axis spectrometers) using single neutron detectors that can be accomplished readily using super mirrors or Heusler crystals. Thorough discussions of this topic and comparisons of these methods to ^{3}He NSFs can be found in [8, 9].

2. Neutron polarizing supermirrors vs. ^{3}He neutron spin filters.

2.1. Polarizing supermirrors

Neutron polarizers/analysers built on polarizing supermirrors (SM) are widely used in neutron scattering instruments, however only for relatively low energy neutrons. The reason for this is that the critical angle of total reflection of neutrons in the direction normal to the surface, θ_c, for most materials is rather small and is about 0.1λ °Å^{-1} maximum, where λ is the neutron wavelength. Using interference multilayer coatings made of thin alternating layers of materials such as Ti, Ni, Fe, or Co of different thicknesses, called supermirrors, one can increase θ_c by the factor *m*. In recent years substantial progress in the manufacturing of polarizing super mirrors has been achieved that allows commercial production of SMs with *m* up to 4.5. An example of the performance of an industrially produced SM with a very high polarizing efficiency of 99 % and a good reflectivity of about 80 % is shown in Fig. 1 [10].

However, in spite of such a high critical angle, still a very large area of SM coating is required for a real multichannel analyzer made of stacks of many mirrors separated by a small space determined by θ_c , mirror length and the beam divergence defined by the incident beam parameters. Moreover, today the flux on the sample for many instruments is increased using focusing guide systems, which further increase the beam divergence. For divergent neutron beams, the angle of incidence on the mirror should be below θ_c for any neutron in the beam, so that the mirror should be installed at the angle $i=\theta_c -\beta$ (2β is the beam divergence).

In addition, SMs suffer from polarizing efficiency losses because of the critical reflection from Si seen

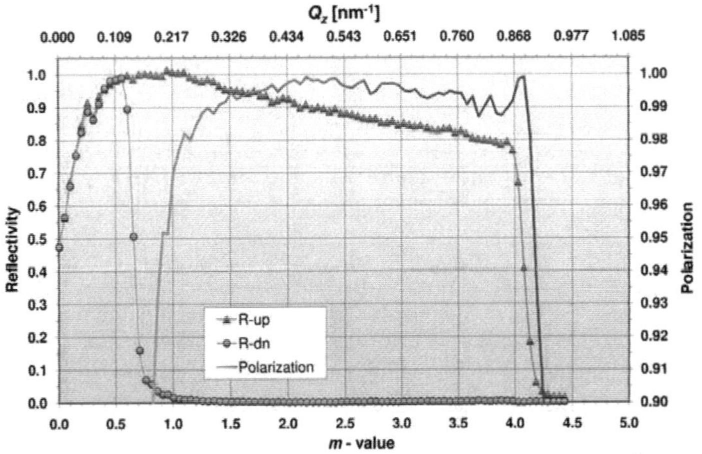

Figure 1. Spin dependent reflectivity and polarization of an Fe/Si polarizing supermirror m=4 [10].

on the left hand side of Fig. 1 for Q_z<0.11 nm^{-1}, where both spin-up and spin-down components are reflected and the beam polarization becomes zero. Practically, the minimal m value that still provides a high polarization of >95 % is $m_{\min}$=0.9, that limits the useful m-range to $\Delta m = m_r$=3.1 for the mirror in Fig. 1. Indeed, for the neutron beam wavelength λ=2 Å the maximal beam divergence (in the direction normal to the SM; the divergence in the plane of the SM can be as large as desired) that corresponds to m_r is $2\beta = \Delta m \lambda \theta_c$ =0.62° (θ_c=0.1°/Å) and the incident angle is $i=\beta+\theta_{\min}$ =0.49°, where $\theta_{\min}=m_{\min}\lambda\theta_c$. Thus to cover the whole solid angle Ω_w of the detector one needs $\Omega_w/2\beta$ SMs of the length L required to overlap a neutron beam of width s. Since $L=s/sin(i)$, then L=117 cm of SMs with m=4 are required to overlap 1 cm of beam width at λ =2 Å. For a large detector on a TOF spectrometer or diffractometer with Ω_w=150° one would need about 280 meters of SMs of a height defined by the vertical scattering angle Ω_h accepted by the detectors and by the sample to analyzer distance. Indeed, if the SM analyzer is midway between the detector and sample for a 2 m detector height, then the total area of SMs amounts to a few hundred square meters. Moreover, as $i \propto \lambda$, this amount is increased inverse proportionally to the neutron wavelength.

While SMs are indispensible for incident beam polarization, where the beam size is limited to the typical size of a neutron guide and the beam divergence is low, and have been successfully used for wide-angle analyzers for cold neutron reflectometers (e.g. NERA [11], HADAS [12], CRISP [13], PRISM [14] etc.) and spectrometers (e.g. D7 [3], DNS [15], etc.),

Journal of Physics: Conference Series **294** (2011) 012013 doi:10.1088/1742-6596/294/1/012013

their use for some of our applications would be difficult. For example, the wide-angle analyzers for the short wavelength neutrons and large area detector of our thermal TOF spectrometer TOPAS, and coverage of large area detectors with sufficient resolution such as our cold neutron SANS, seem to be best served by an approach using ^{3}He NSFs. For spin flip scattering that is a small fraction of the total scattering, arguments have been made for the use of ^{3}He polarization analysis on fundamental bases of the obtained data quality achievable with ^{3}He NSFs, however this is a rather specific discussion that has been covered in prior literature [9] which we will not discuss further here.

2.2. ^{3}He neutron spin filters

The operation of ^{3}He NSFs is in a way similar to that of SM neutron polarizers placed in a transmitting geometry: both are filtering out the undesirable spin component, the only difference is that it is absorbed in the first case and reflected out of the beam in the second case. Both of them allow for the undisturbed propagation of the neutron beam – a feature that is tremendously important for high angular resolution applications such as small-angle neutron scattering and reflectometry. Another advantage of the ^{3}He NSFs is that they allow for the precise neutron beam polarization knowledge (>99.9 %) that is of a great importance for an accurate determination of small spin-flip signals [16, 17].

In contrast to SMs, ^{3}He NSFs are not subject to the above-described geometrical scaling with the decrease of the neutron wavelength. Their efficiency is determined by the filter opacity that can be controlled by the polarized ^{3}He gas pressure so that they are rather compact even when used for thermal and hot neutrons (i.e. for λ<1 Å). There are some practical considerations such as the safety limits of cells pressurized to over atmospheric pressure and reduction of ^{3}He T_1 lifetime due to increased dipole-dipole interactions to be discussed below. However, because of their compactness compared to SMs, ^{3}He NSFs can provide certain practical advantages for the high solid angle coverage at a lower price.

On the other hand, ^{3}He NSFs are not free from disadvantages. The main drawback is the time decay of the ^{3}He polarization P_{He} in the cell which leads to a lowering of the neutron beam polarization P_n and transmission over time if the polarized ^{3}He gas is not refreshed. This decay is caused by some intrinsic reasons such as dipole-dipole interaction of the ^{3}He atoms, and by the space inhomogenity of the magnetic holding field [18-20]. Further, interactions with the cell's walls also cause ^{3}He relaxation in most common instances. The resulting achieved ^{3}He lifetime is the sum of the three aforementioned sources: interactions with the cell's wall, field gradients and dipole-dipole interactions can be respectively written as

$$\frac{1}{T_1} = \frac{1}{T_{cellwall}} + \frac{7000}{p}\left(\frac{\partial B/\partial r}{B}\right)^2 + \frac{p}{750} \tag{3}$$

Here T_1 is in hours, p is the ^{3}He pressure in bar and the quantity squared in the second term is the transverse magnetic field gradient in cm^{-1}. Since the second and third terms of this equation have the opposite dependence on pressure, in the limit where the magnitude of the magnetic term is comparable to the dipole-dipole term the pressure can actually be used to optimize the cell lifetime. 100 hour lifetimes are generally considered to be the lower acceptable limit because at this level one would have to refresh the ^{3}He polarization on a daily basis in order to retain reasonably good performance in terms of P_n and neutron transmission [21]. Stray magnetic fields from external magnets reduce performance and, beyond certain limits because of the quadratic dependence of this term, can result in rapid losses of the neutron beam polarization and transmission if care is not taken over the shielding from such sources.

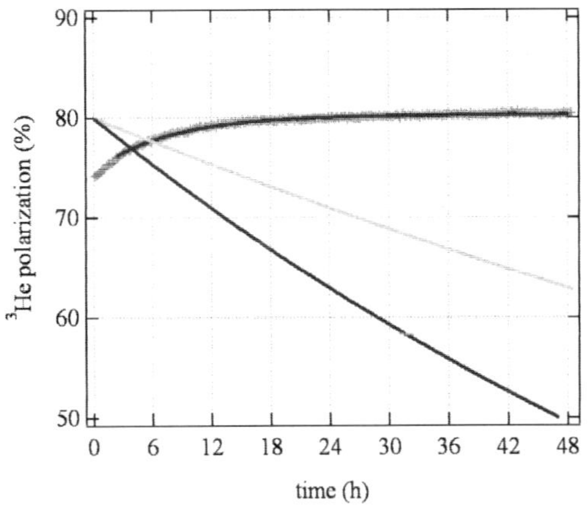

Figure 2. Time-dependence of the ^{3}He polarization by in-situ SEOP method [23]. The green/blue lines indicates the ^{3}He NSF performance if the same cell were polarized to the same level and then were allowed to undergo normal T_1 nuclear spin decay of the polarization with a 200/100 hour time constant respectively.

Indeed, to boost the performance of ^{3}He NFSs reduction of the negative factors is required. Clearly one should attempt to reach the longest life-times for the polarized ^{3}He gas, so that changes of the neutron beam polarization during the neutron scattering experiment will not be as important and so that the frequency of refreshing the polarization is minimized. However the ultimate solution of the problem is to preserve the ^{3}He polarization at a high and constant level. This can be achieved using continuous optical pumping of the ^{3}He, i.e. matching the pumping rate with the rate of ^{3}He polarization losses such that the rate balance results in a high polarization at constant level [22]. The so called spin-exchange optical pumping method (SEOP) in principle allows one to keep a high polarization for an essentially unlimited time [23]. Recently we achieved a very high on-beam time averaged ^{3}He polarization of above 80 % for over a day (Fig. 2), where the time limit was actually imposed by the experiment duration and not the equipment used. In further testing using a larger cell a good ^{3}He polarization of 73 % was obtained and maintained for about 80 hours. The measurement which included the use of a sample magnet at up to 1 T and frequent reversal of the direction of ^{3}He polarization by the adiabatic fast passage method did not have any effects on the ^{3}He polarization or stability. A more detailed discussion about the comparative performance of ^{3}He NSFs with constant and time decaying polarization can be found in [24].

We would like to mention that one can overcome in some way the time decay problem although by a regular refilling of the cell with freshly polarized ^{3}He gas or by exchanging the cell with a freshly polarized one, thus keeping the averaged polarization over the experiment duration on a reasonable level with the time decaying ^{3}He NSF method. This results in a kind of regular (often daily) maintenance of the ^{3}He NSF that requires substantial manpower. On-beam polarization in contrast allows a practically maintenance-free operation. Furthermore, neutron experiments performed with ^{3}He polarization held at a constant level using the in-situ SEOP method do not need time dependent polarization corrections to analyze their data and will have the highest obtainable time averaged neutron performance for a given set of experimental conditions. However, for certain applications using banana-shaped cells and/or variable magnetic field directions [25-27] it is not apparent how in-situ polarization could be done successfully, thus a local filling approach, or routine change of cells is envisioned.

Journal of Physics: Conference Series **294** (2011) 012013 doi:10.1088/1742-6596/294/1/012013

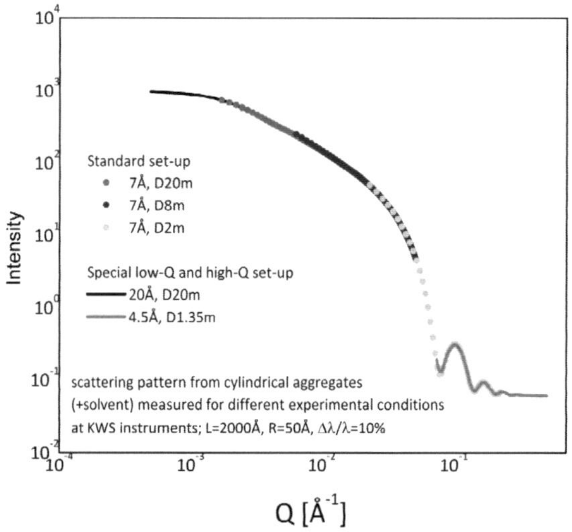

Figure 3. An example of the SANS scattering curve from elongated cylindrical aggregates (length 2000 Å and radius 50 Å). The curve was measured by parts at different experimental conditions (the sample-detector distance and the wavelength).

3. ^{3}He polarization analyzers for different neutron instruments at JCNS

Now we will consider instrument requests and possible design parameters of polarization analyzers for a few different classes of neutron scattering instruments, namely SANS, reflectometry and wide angle TOF spectrometry.

3.1 Small-angle neutron scattering (SANS)

One can define two limiting cases for the operation of a SANS diffractomer: the highest Q-resolution limit and the highest Q-range limit (here Q is the momentum transfer). In the first case the aim is to achieve the minimal possible Q_{min} that allows for studies of large scale structures. As the Q_{min} of the detector part of the SANS machine is defined by the ratio of the detector pixel size to the sample-detector distance L, this requires the latter to be as large as possible. Therefore SANS instruments are usually very long, most commonly about 20 m, or even up to 40 m (D11 at the ILL [28]) that allows for pushing Q_{min} down to about 10^{-4} Å^{-1}. However, in this case the Q-range available for the measurements that is defined as the ratio of detector size H_D to L (i.e H_D/L) is rather small.

In the second case, the detector is placed very close to the sample, with L down to 1 m, so that the Q-resolution of the detector part of the SANS machine is very low, however the Q-range is extended to about 1 Å^{-1}. In general, the SANS diffraction spectrum is usually taken by three separate measurements covering large Q area with low resolution ($L \rightarrow L_{min}$, $\lambda \rightarrow \lambda_{min}$), than exploring low Q-range with high Q-resolution ($L \rightarrow L_{max}$, $\lambda \rightarrow \lambda_{max}$) and then one measurement with intermediate resolution (L=7-8 m, $\lambda \approx 2\lambda_{min}$) that overlaps with both previous measurements. In this way one covers up to 3-4 orders in Q with an optimal resolution over the different parts of the SANS curve (Fig. 3).

Now let us consider the problem of polarization analysis using a ^{3}He NSF on such a setup. Principally one can use the same ^{3}He NSF for all 3 measurements, however the requirements to the quality of optical cells are quite different, if not opposite, in the limit cases. Imperfections of the cell surface may play a role as a scattering object so that the scattering from the cell material will be a source of background that can overwhelm a small desired signal. Indeed, the cell walls exposed to the neutron beam should have tremendously good surface to minimize the undesired scattering from the surface roughness.

3.1.1 Cell material

The most suitable material for ^{3}He SEOP cells is GE180 glass produced by General Electric Co [29]. Such cells have been shown to provide the longest lifetime of the polarized ^{3}He and are suitable for neutron use because GE180 does not contain boron, which is a strong neutron absorber, like most common glasses do. To check the feasibly of a typical GE180 glass cell for various types of neutron scattering experiments, we carried out a SANS experiment aiming to study the surface scattering from such a cell's walls. To avoid any artefacts the empty cell blown from GE180 glass, identical to those used for SEOP to polarize ^{3}He gas, was placed in the SANS instrument KWS1 [30] of the JCNS that allows for a common vacuum flight path over the whole setup, so that there is no other material, besides the sample, in the neutron beam. Indeed, the scattering curve represents solely the scattering on the cell's walls without any extra contributions and therefore requires no corrections.

As one can see from the Fig. 4, the cell made of GE180 glass as blown is showing excessive scattering for $Q<0.01$ Å^{-1} and therefore shouldn't be used in this Q-range if the cell is relatively close to the sample and thus far from the detector. However, for $Q>0.01$ Å^{-1} the scattering from the cell is quite small. As the reference level we choose the scattering from 1mm of D_2O because it is used in standard buffer solutions for soft matter samples and represents the lowest scattering power of samples to be measured, (noting most samples have a thickness of 2-5 mm depending on concentration). As one can see, the scattering from the cell is even slightly lower than that from 1mm D_2O for $Q>0.01$ Å^{-1} therefore any contributions to the scattered intensity in this regime will be small or negligible.

But how about for $Q<0.01$ Å^{-1}, especially for the smallest Q of about 10^{-4} Å^{-1}? After tests of different materials with different surface treatments, we found that Si wafers that underwent a special chemically-mechanically polishing are providing the best result: their roughness is sufficiently small and as one can see from the result of the experiment (Fig. 4) the level of neutron scattering is so low that it may be considered to be negligible within the error bars of this measurement. While Si-windowed cells have been used for off-line polarized ^{3}He NSFs when the gas is polarized outside of the cell, to date no cells employing Si windows have been successfully made to employ direct optical pumping of the cell using the SEOP method which requires heating of the cell commonly up to 190°C or higher. Si-windowed cells have been produced for gas polarized outside of the neutron target cell using either MEOP [31], in which the gas is polarized remotely and transferred to the cell through a valve, or with SEOP wherein the Si-windowed cell was connected to a SEOP suitable GE180 cell via a thin polarization transfer tube to allow exchange of ^{3}He gas between the two volumes [32]. For the case of the SEOP cell more development is needed before application to neutron scattering. For very high Q-resolution, it should not be overlooked that a polarized ^{3}He cell could be placed near the detector, thus enabling use of normal GE180 cells. For example, SANS diffractometers KWS3 or KWS2 at the JCNS in the highest Q-resolution mode (Q~10^{-4}) do not use the standard large area detector, but rather 9 cm in diameter scintillation detectors, which could be covered with one ^{3}He NSF cell, of a common diameter for either the SEOP or MEOP method.

Journal of Physics: Conference Series **294** (2011) 012013 doi:10.1088/1742-6596/294/1/012013

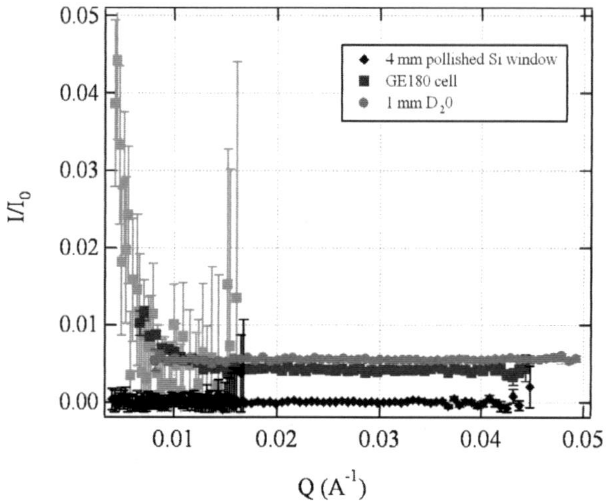

Figure 4. SANS spectra for different materials to be used as windows for neutron beams.

3.1.2 Cell size

In Fig. 5 the placement of a ^{3}He NSF analyzer in a SANS instrument is shown for two limiting cases. As one can see in the first case of a large Q-range (a) the cell diameter should be rather large to span the whole scattered beam. Assuming that cell is installed at distance l from the sample the required cell size D_{cell} is

$$D_{cell} = d_{sample} + H_D \frac{l}{L_1} \tag{4}$$

so that D_{cell} = 5 cm for a typical sample size d_{sample}=1 cm, detector size H_D=1 m, l=5 cm and the sample-detector distance L_1=1.3 m. (Here we assumed that the sample is put close to the cell's wall which is only possible in the case of samples that do not create field gradients at the ^{3}He cell position (which is an evident case for soft matter or biological samples). In this case there is also no strict request to the quality of the cell's walls and therefore GE180 cells as blown can be used without any complications.

For the other case of high Q-resolution (b), L_2 is about 20 m, so that for the same parameters H_D and l, the cell diameter can be much smaller and be practically equal to the sample size, i.e. about 1cm. However, in this case there is a strong request to the quality of the cell walls' surfaces that as mentioned could be fulfilled by the use of windows made of chemically polished Si.

Certainly one can try to make a cell that will satisfy both cases, however it seems to us to be most appropriate to use a standard GE180 blown cell with a good lifetime and a small Si windowed cell polarized with a diffusion tube to be discussed below for the second case. A third possible case presented at the end of section 3.1.1 in which for very high Q-resolution, low Q-range SANS, a cell could be placed close to the detector in which case small amounts of scattering from the cell walls would be unimportant.

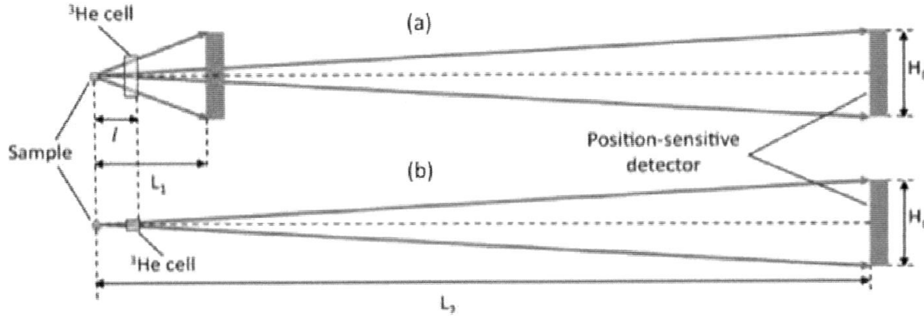

Figure 5. Placement of NSF analyzers in the a SANS instrument for two limiting cases: (a) a large Q-range and a low Q-resolution; (b) a small Q-range and a high Q-resolution.

3.1.3 Cell polarizing efficiency

Another problem is that the maximal balance of neutron transmission and neutron polarizing power of a ^{3}He NSF is achieved when the opacity, i.e. the product of the gas pressure in bar, the cell thickness in cm and neutron wavelength in angstroms is about 30. At opacities much below 30 for commonly obtained ^{3}He polarizations, the polarising efficiency of the ^{3}He NSF becomes unacceptably low, i.e. <90 %. For opacities to far above the optimal opacity, again for commonly obtained ^{3}He polarizations, the cell transmission becomes unacceptably low. Indeed, the cell optimized for short wavelength neutrons for the case of the low Q limit (a) (see Fig.5), would be too opaque if used for long wavelength neutrons required in the case (b) resulting in inacceptable intensity losses. In the opposite case, if the cell optimized for the high Q-resolution case (b) (i.e. for long wavelength neutrons) is used for the case (a), then the cell opacity for the case (a) (i.e. for short wavelength neutrons) will be too low resulting in inacceptable poor neutron beam polarization. A full discussion of optimal ^{3}He NSF opacities is somewhat sample dependent and is discussed in detail in ref. [9].

A possible solution that we are working on is a double-chambered cell made of two different cells connected by a diffusion tube to transfer polarized ^{3}He gas between them. One of them will be a large diameter GE180 cell with its thickness and diameter optimized for the case (a), another - a much smaller Si-windowed cell with a smaller thickness and diameter optimized for the low Q limit. The larger GE180 chamber would be heated in an oven and polarized like a normal SEOP cell whereas the smaller Si-window chamber would be outside of the oven, remaining at room temperature, and be passively polarized via. gas exchange with the GE180 cell. In this case, both chambers of the double cell will be also optimized from the point of view of the allowable scattering from their walls'. This idea is in some ways an extension of an idea formed for fundamental physics beam line, and a double cell was created for this purpose, however in this case it was never planed that the GE180 cell would be used for an NSF as well. The cell in that work showed a good T_1 relaxation time for the ^{3}He implying on should be able to polarize it well, however more testing is needed [33]. Therefore more work needs to be done on the principle of double chambered Si-windowed plus GE180 cells before it can be practically applied.

An example of the possible performance of such device filled with ^{3}He polarized at 70 % is presented in Fig. 6. In this graphic we assumed a rather modest polarization compared to those obtained in normal one-chamber GE180 SEOP cells. Increasing the ^{3}He polarization beyond this value would serve to further increase the allowable bandwidth of each portion of such a double chambered cell. As one can see a high dynamic range for the polarizing power over a large lambda range can be realised, which would allow for high-quality measurements for all three experimental conditions shown in Fig.6.

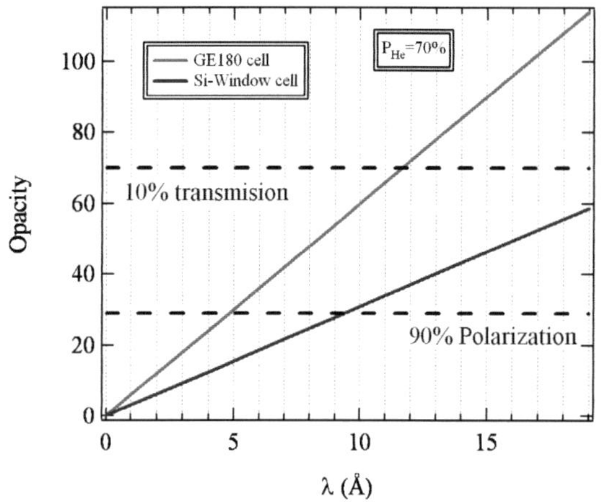

Figure 6. Example assuming P_{He}=70% and a double chambered cell with a large diameter GE180 cell optimized for the high Q limit connected by a diffusion tube to a much smaller Si-windowed cell optimized for the low Q limit. In principle a double cell such as this could be effectively used starting at 5 Å and retain good performance at well over 20 Å.

3.2 Reflectometry

In polarized neutron reflectometry (PNR) our scientific cases of interest are mostly studies of magnetic nanostructures, as MARIA will also measure polarized GISANS (Grazing Incidence SANS), and MBE (molecular beam epitaxy) grown layered systems with strong interlayer exchange, both of which require high fields on the order of 1 T or more to saturate the magnetization [34, 35]. Such magnets generate fringe magnetic fields that can result in magnetic field gradients over the cell and in turn lead to a serious reduction in the lifetime of the polarized ^{3}He. To reduce the influence of such fringe fields, one should try to keep the ^{3}He cell far enough away from the magnet and/or design effective shielding of the sample magnet field at the ^{3}He NSF position. This however results in a significant increase of the cell size: e.g. the magnetic reflectometer MARIA at the JCNS [36] employs a large position sensitive detector with H_D=0.4 m and L_I=2 m, so that the diameter of the cell should be about 14 cm when placed outside the sample magnet at a distance of l=60 cm from the sample (d_{sample}=1 cm) where the 60 cm represents the minimum distance allowed by our current mu-metal shielded magnetic cavity design [13]. For the scattering measured on a typical GE180 cell we would expect the minimum spot size resolvable due to the scattering from the cell to be $<$ 2 mm at 4.5 Å. As the detector resolution for MARIA is also about 2 mm we do not expect the use of a GE180 cell to appreciably degrade detector resolution. However, for the optical pumping of such a large cell, the hybrid K-Rb SEOP method should be used in order to maximize the efficiency of the lasers used in order to effectively polarize the large 1.5 l volume [37].

3.3 TOF spectrometers and diffractometers

The same considerations as for PNR are valid also for TOF spectrometers and diffractomers requiring ultra wide-angle analyzers as in most studies of magnetic materials or strongly electron-correlated systems, strong magnetic fields are applied. As an example, for the thermal neutron spectrometer TOPAS being constructed by our group, the JCNS at FRM II, a compensated 5 T magnet will be used that has been designed to give a fringe field below

0.1 mT at maximum field at a distance of 60 cm from its center. If the NSF is placed at this position (60 cm), then the volume of a banana-shaped ^{3}He cell with a 10 cm thickness and a sufficient height to accept the entire scattered beam would be about 50 litters. To be an effective analyser for thermal neutrons with λ=1 Å the ^{3}He pressure in such a cell should be about 3 bar, so that about 150 bar·litres of polarized ^{3}He will be necessary to fill such a volume.

Though in principle the magnet can either be shielded or further compensated to give acceptable T_1 lifetimes, nevertheless the open geometry of such a large magnetic system will hardly provide an ideal magnetic environment for the ^{3}He NSF. This will certainly place limits on the maximum attainable lifetimes of the ^{3}He polarization. Therefore, a frequent and probably daily refilling of the cell is envisioned. As the required amounts of polarized gas, about 100 bar·litre/day, cannot be produced by any current or planned metastable exchange optical pumping (MEOP) stations to our knowledge, a high output SEOP pumping station is being considered as such a system is currently under development (see e.g. [38]). We do not preclude that eventually higher production rate MEOP systems could become possible as much fundamental work is being done on this area [39, 40]. Consequently any future system cable of polarizing the required volumes would be suitable for this application.

4. Application of polarization analysis in soft matter studies

As was already mentioned in the introduction, using longitudinal or XYZ polarization analysis one can separate different contributions to the neutron scattering cross section [41, 42]. When applied to soft matter studies, the power of both these methods is in the possibility to separate coherent and incoherent contributions. The incoherent scattering in this case is caused mainly by hydrogen atoms which always constitute a significant part of atoms in any polymer or biological molecule with a very large incoherent cross section of 80 barn per atom. The latter should be compared with coherent cross sections of the other most abundant constituents of such molecules: 5.5 barn for carbon, 11 barn for nitrogen and 4.2 barn for oxygen per atom. Therefore, the hydrogen atoms in polymer and biological molecules are strong intrinsic sources of neutron background at the detector. Because the partial incoherent cross section of hydrogen in a molecule is usually much larger than the partial coherent cross sections of the other elements in the same molecule, the intensity of the isotropic incoherent scattering in certain Q regimes is much higher than the intensity of the anisotropic coherent scattering. In other words, the coherent scattering is centered around Q=0 and decreasing with a power law Q-dependence ranging from Q^{1} to Q^{4} (corresponding to the two limits of infinitely long rods and hard spheres respectively) whereas the incoherent scattering is flat such that at high enough Q the incoherent scattering will dominate for any sample. This means that the useful structural information in the diffraction pattern at high Q is covered to various degrees by an overwhelming isotropic incoherent scattering background and the resulting signal-to-noise ratio in the extracted coherent scattering signal is very low.

Moreover, because in this case the incoherent background is caused by the sample itself, the normal differential approach which is to perform two measurements - one with the sample and one without, then subtract these two patterns and to consider their difference as the scattering from the sample is not applicable at all. Straightforward calculations of intensity caused by the incoherent scattering from the sample are not useful because of the multiple incoherent scattering processes that results in the significant, sample geometry dependent, increase of the incoherent cross-section [43].

Additionally, because systematic errors in the absolute calibration of the measured scattered intensity in the units of scattering length (cm^{-1}) are not less than 10 %, the subtraction of the analytically calculated incoherent background is not possible. However, it is not only difficult to obtain the relevant coherent scattering from the sample, in addition measurements with a strong incoherent background result in an ambiguity of the fit parameters determined from the scattering curve, which we illustrate graphically in Fig. 7. A thorough discussion of the problems in the determinations of protein structures arising from incorrect subtraction of the "baseline" scattering can be found in ref [44]. Here the scattering curve from a protonated structure is presented in a Porod plot (the log(I)-log(Q) presentation). The architecture is analyzed using the empirical Beaucage model [45, 46] that is characterized by three fitting parameters: the Guinier scaling factor G, the radius of gyration R_g and the Porod exponent d. However, because of the unavoidable incoherent background the fourth fitting parameter, the Q-independent

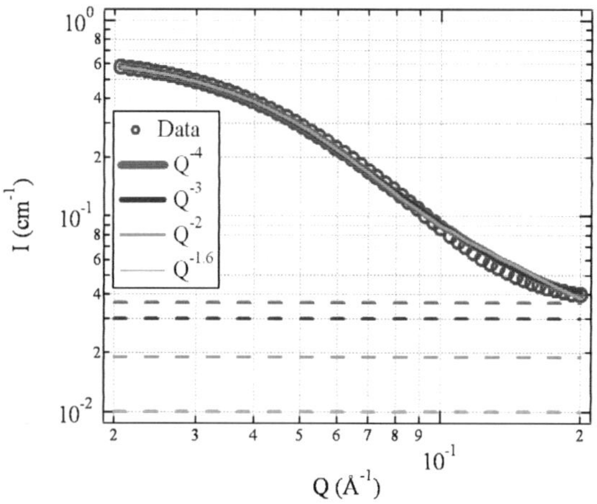

Figure 7. Beacauge fits to the scattering curve obtained with unpolarized neutrons (open grey circles). The four solid lines, which are nearly impossible to distinguish because they overlap, are fitted with an exponent d equal to -4 (red), -3 (blue), -2 (green) and -1.6 (orange) with different levels of background (the dotted lines of the same color) [47].

background intensity I_{bckg}, must be introduced to describe the scattering pattern over the entire Q-range covered in the measurement. This results in a strong correlation between this parameter with the power exponent d that delivers the morphology of the structures. Therefore, changing the assumed level of background I_{bckg} one can fit the scattering curve with very different power laws, with d ranging from -1.6 to -4 (see Fig. 7), that makes an unambiguous determination of the sample morphology impossible. This is especially important in SANS studies of biological structures, where a proper and accurate subtraction of the solvent contribution and the determination of the incoherent background level from the sample becomes particularly critical in the Q-range above 0.1 Å^{-1}.

A standard solution of this problem is the reduction of the incoherent background by the isotopic substitution of hydrogen atoms with deuterium, which is chemically equivalent but has an incoherent cross section of only 2 barn and a coherent cross sections of 5.6 barn per atom. However, the procedure of isotopic substitution is rather complicated and time consuming, so that not all substances can be prepared in the deuterated form in an amount sufficient for a neutron scattering experiment. Indeed, in some neutron research centers entire laboratories have been created entirely dedicated to the problem of sample deuteration.

As mentioned above neutron polarization analysis is another way to solve this problem. Using a polarized incident neutron beam and analyzing the spin state of scattered neutrons one can distinguish between the coherent and incoherent components [41], effectively raising the signal-to-noise ratio and eliminating the ambiguity from incoherent background. The possibilities of this technique with the aide of a ^{3}He NSF was demonstrated in [48]. Recently we applied it in an experiment carried out on the SANS diffractometer KWS2 of JCNS at FRM II in Garching (Germany). We studied the structure of lactoferrin (a protein with antimicrobial activity) at 1 % protein concentration in a D_2O buffer solution (pH7). As a non-deuterated protein it is the source of a strong and unavoidable incoherent background.

To provide a wide angle for the polarization analysis the sample was placed close to a polarized ^{3}He NSF allowing it to span a solid angle of about 15 ° corresponding to $Q<0.35$ Å^{-1}. The incident neutron beam was polarized by a transmission super mirror polarizer (the beam polarization was 96.6 % at 4.5 Å). A Mezei-style spin flipper was used before the sample to flip the incident spin with an efficiency of about 97 %. The ^{3}He cell was polarized in the JCNS SEOP laboratory up to the starting polarization of 72 %. The cell was the transported to KWS2 and placed in a long shielded, end-compensated solenoid system with a high field homogeneity. A picture of the experimental installation is shown in Fig. 8.

Figure 8. Solenoid system of the polarized neutron SANS. The sample is placed inside the solenoid system a few cm before the ^{3}He spin filter cell. An overlap of magnetic fields of the polarizer/flipper on the left and the solenoid system allows for the adiabatic rotation of the neutron spin when propagating from the vertical guiding magnetic field into the horizontal magnetic field of the solenoid system.

This ^{3}He NSF provided a high polarization efficiency of 95 % at the beginning of the experiments and remained over 93 % after over 36 hours of measurements. The setup allowed us to obtain a lifetime T_1 of about 430 hours, which is only slightly lower than the lifetime of this cell in ideal conditions, which is 550 hours (see [47] for more details). The level of lifetime performance achieved here would allow us to operate for 4 to 5 days on a single polarization of the cell with a good neutron performance however the data would need to be corrected for the time-decay as stated earlier. Please note that these measurements do not require a flipper after the sample. As the samples have no magnetic scattering all spin-flip scattering events can be interpreted as the result of incoherent scattering processes. A

Journal of Physics: Conference Series **294** (2011) 012013 doi:10.1088/1742-6596/294/1/012013

detailed description of this method is given in [41, 42].

Errors could occur due to significant changes in the energy of the neutrons transmitted through the sample from inelastic scattering, which will increase the neutron energy, i.e. decrease λ, the neutron wavelength, of a portion of the transmitted neutron spectrum [49]. Because of the neutron-energy dependent efficiencies of detectors and the ^{3}He NSF's energy dependent polarization analyzing efficiency, a $\tanh(\lambda\Theta P_{He})$ dependence, and transmission, a $\exp(\lambda\Theta)\cosh(\lambda\Theta P_{He})$ dependence, inelastic scattering could lead to errors in the data analysis. Here Θ is the ^{3}He NSF cell opacity or density-length-cross section product, and P_{He} is the ^{3}He polarization. For example, in the limit where the polarization analyzing efficiency is about one for all expected λ, and a "ideal" neutron detector efficiency, the inelastically up-scattered neutrons will be weighted roughly exponentially as a function of their wavelength. However since the sample is actually a small fraction of the sample plus solvent, and the solvent scattering signal will be measured in both spin parallel and anti-parallel states and subtracted off during data analysis, we should only worry about the effects on the data arising from the inelastic scattering of the sample itself. Thus one should have a reasonable knowledge of the inelastic spectrum from the sample to determine whether or not it will be in the limit where a correction must be made. We will consider these effects carefully in future work. Using time of flight analysis we will study the inelastic spectrums from typical and extreme case samples. Pending the outcomes of this work we will work to develop the required methods and/or corrections for proper data treatment. The data analysis should also account for effects of multiple scattering on the PA, a discussion of which is given in Ref. [48]. A more detailed discussion of the entire SANS data analysis for PA with ^{3}He is not a subject of the current article, but may be considered for future work.

A plot of the data obtained is shown in Fig. 9. The black circles are the result of standard measurements with an unpolarized neutron beam (raw data thanks to C. Sill, Forschungszentrum Jülich GmbH). The data are solvent-corrected using a standard approach for the treatment of SANS data from solutions [50]. The red circles are obtained after the data treatment using a polarized incident beam with ^{3}He polarization analysis. The

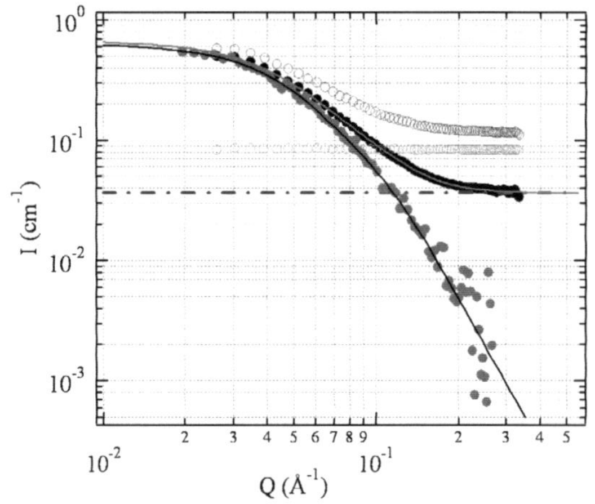

Figure 9. SANS patterns from 1 % protein lactoferrin in D_2O buffer (pH7) [40] solution measured with unpolarized neutrons (black symbol) and polarized neutrons and polarization analysis (red symbols). The red solid line is guided by eye. The Q-dependence at high Q can be estimated as $Q^{-3.85}$.

comparison of these two data sets show the ability to measure the coherent portion of the scattering that is on the order of 100 times below the level of the samples own incoherent scattering, making an unambiguous fit procedure possible. We do not claim that this technique should replace current well-developed techniques to obtain information on the coherent scattering intensity, such as contrast variation methods or to lower the level of incoherent scattering through deuteration, however we feel it will be very useful in difficult cases. The utility of such an approach has been shown in [32] for example.

5. Results and conclusions

Polarization analysis is an imperative technique for polarized neutron scattering when the full information about the vector magnetization in the sample is required. Thus for the detailed understanding of physical properties of molecular magnets, new superconductors, spin electronic and magnetic nanostructures, as well as the self-organization of magnetic nanostructures PA is a must for modern neutron instrumentation. Further it was shown that there are experimental areas of polarized neutron scattering where the polarization analysis using polarized ^{3}He spin filters is beneficial, compared to polarizing supermirror analyzers.

In its simplest 1-dimensional version, polarization analysis allows for the separation of coherent and incoherent scattering that could become a useful technique for studies of protonated biological objects whose hydrogen atoms are the source of a strong intrinsic and principally unavoidable background. It was demonstrated in a relevant SANS experiment aimed for the study of the structure of the prorogated protein lactoferrin that the signal to noise ratio can be effectively increased by about two orders of magnitude, thus allowing for the unambiguous interpretation of the experimental data that was impossible without polarization analysis.

Acknowledgements

We are thankful to C. Sill for providing us with the lactoferrin sample, M.-S. Appavou, V. Ossovij, A. Nebel, S. Staringer and Th. Kohnke for support during the experiment. We also point out S. Mattauch's continual involvement with our polarized ^{3}He efforts especially with respect to reflectometry application. Lastly we thank the reviewer for pointing out the inelastic scattering issue which we will study further with respect to PA in soft matter studies.

References

[1] Rekveldt M Th 1973 *Zeitscrift fur Physik* **259** 391
[2] Rekveldt M Th 1976 *J. of Magnetism and Magnetic Materials* **1** 342
[3] Stewart J R, Deen P P, Andersen K H, Schober H, Barthelemy J F, Hillier J M, Murani A P, Hayes T and Lindenau B 2009 *J. of App. Crystalography* **42** 69-84
[4] Voigt J, Babcock E, Bruckel Th 2010 *J. Phys. Conference Series* **211** 012032
[5] http://www.isis.stfc.ac.uk/instruments/wish/
[6] Ollivier J, Mutka H and Didier L 2010 *Neutron News* **21 issue 2** 22
[7] http://www.jcns.info/jcns_instruments/
[8] Cussen L D, Goossens D J and Hicks T J 2000 *Nuclear Instruments and Methods A* **440** 409-20
[9] Cussen L D and Goossens D J 2002 *Nuclear Instruments and Methods A* **490** 316–333
[10] http://www.swissneutronics.com
[11] Solina D, Lott D, Tietze U, Frank O, Leiner V and Schreyer A 2006 *Physica B: Condensed Matter* **385-86** 1167-9

[12] Rucker U, Bergs W, Alefeld B, Kentzinger E and Bruckel Th 2001 *Physica B: Condensed Matter* **297** 140-2
[13] Nunez V, Boothroyd A T, Reynolds J, Penfold J, Langridge S, Bucknall D G, Böni P, Clemens D and Senthil Kumar M 1997 *Physica B: Condensed Matter* **241-243**, 148-50
[14] Fermona C, Ott F, Legoffa G, Glättlia H and Wintenberger V 2000 *Physica B-Condensed Matter* **283** 372-5
[15] Schweika W and Böni P 2001 *Physica B: Condensed Matter* **297** 155
[16] Chupp T E, Coulter K P, Kandes M, Sharma M, Smith T B, Jones G, Chen W C, Gentile T R, Rich D R, Lauss B, Gericke M T, Gillis R C, Page S A, Bowman J D, Penttila S I, Wilburn W S, Dabaghyan M, Hersman F W and Mason M 2007 *Nuclear Instruments and Methods A* **574** 500-9
[17] Zimmer O, Muller T M, Hautle P, Heil W and Humblot H 1999 *Phys. Lett. B* **455** 62-8
[18] Newbury N R, Barton A S, Bogorad P, Cates G D, Gatzke M, Mabuchi H and Saam B 1993 *Phys. Rev. A* **48** 4411
[19] Dénigre A, Heil W, Otten E W, Wolf M, Kremer R K, and Simon A 2006 *European Phys. J. D* **38** 439-43
[20] Hasson K C, Cates G D, Lerman K, Bogorad P and Happer W 1990 *Phys. Rev. A* **41** 3672-88
[21] Babcock E and Ioffe A 2011 *IOP journal of physics conference series, this volume*
[22] Boag S, Babcock E, Andersen K H, Becker M, Charlton T R, Chen W C, Dalgliesh R M, Elmore S D, Frost. C D, Gentile T R, Anton R L, Parnell S R, Petoukhov A K, Skoda M W A and Soldner T 2009 *Physica B: Condensed Matter* **404** 2569-2
[23] Babcock E, Mattauch S and Ioffe A 2011 *Nuclear Instruments and Methods A* **625** 43-6
[24] Babcock E and Ioffe A 2011 *Physica B: Condensed Matter* (corrected proof, available online 5 November 2010) ISSN 0921-4526, DOI: 10.1016/j.physb.2010.10.088
[25] Andersen K H, Jullien D, Petoukhov A K, Mouveau P, Bordenave F, Thomas F and Babcock E 2009 *Physica B: Condensed Matter* **404**, 2652-4
[26] Stewart J R, Andersen K H, Babcock E, Frost C D, Hiess A, Jullien D, Stride J A, Barthelemy J F, Marchal F, Murani A P, Mutka H and Schober H 2006 *Physica B: Condensed Matter* **385-86** 1142-5
[27] Chen W C, Erwin R, McIver III J W, Watson S, Fu C B, Gentile T R, Borchers J A, Lynn J W and Jones G L 2009 *Physica B: Condensed Matter* **404**, 2663-6
[28] http://www.ill.eu/instruments-support/instruments-groups/instruments/d11/
[29] GE Lighting Component Sales, Bldg. 315D, 1975 Noble Rd., Cleveland, OH 44117.
[30] http://www.jcns.info/jcns_kws1/
[31] Petoukhov A K, Andersen K H, Jullien D, Babcock E, Chastagnier J, Chung R, Humblot H, Lelievre-Berna E, Tasset F, Radu F, Wolff M and Zabel H 2006 *Physica B: Condensed Matter* **385-86** 1146-8
[32] Gaspar A M, Busch S, Appavou M S, Haeussler W, Georgii R, Su X Y and Doster W 2010 *Biochimica et Biophysica ACTA-Proteins and Proteomics* **1804** 76-82
[33] Babcock E, Boag S, Andersen K H, Becker M, Beecham C, Bordenave F, Chastagnier J, Chen W C, Chung R, Chupp T E, Elmore S, Fouilloux P, Gentile T E, Jullien D, Lelievre-Berna E, Mouveau P, Petoukhov A K, Revert M and T. Soldner 2009 *Physica B: Condensed Matter* **404** 2655-8
[34] de Vries J J, Kohlhepp J, den Broeder F J A, Coehoorn R, Jungblut R, Reinders A and deJonge W J M 1997 *Phys. Rev. Lett.* **78** 3023-6
[35] Krycka K L, Booth R A, Hogg C R, Ijiri Y, Borchers J A, Chen W C, Watson S M, Laver M, Gentile T R, Dedon L R, Harris S, Rhyne J J and Majetich S A 2010 *Phys.*

Rev. Lett. **104** 207203
[36] http://www.jcns.info/jcns_maria/
[37] Babcock E, Nelson I, Kadlecek S, Driehuys B, Anderson L W, Hersman F W and Walker T G 2003 *Phys. Rev. Lett.* **91**, 123003
[38] http://xemed.com/products/hyperpolarizers/helibox_z100/
[39] Batz M, Nacher P-J and Tastevin G 2011 *IOP journal of physics conference series, this volume*
[40] Mrozik C, Endner O, Hauke C, Heil W, Karpuk S, Klemmer J and Otten E W 2011 *IOP journal of physics conference series, this volume*
[41] Moon R M, Riste T and Koehler W C 1969 *Phys. Rev.* **181** 920–31
[42] Schaerpf O and Capellmann H 1993 *Physica Status Solidi A* **135** 359-79
[43] Brulet A, Lairez D, Lapp A and Cotton J P 2007 *J.of Appl. Crystallography* **40** 165
[44] Rubinson K A, Stanley C and Krueger S 2008 *J. of Appl. Crystallography* **41** 456-65
[45] Beaucage G 1995 *J. of Appl.Crystalography* **28** 717
[46] Hammouda B 2010 *J. of Appl. Crystalography* **43** 1474
[47] Babcock E, Ioffe A, Radulescu A, and Pipich V 2010 arXiv:1006.4418v1
[48] Gentile T R, Jones G L, Thompson A K, Barker J, Glinka C J, Hammouda B and Lynn J W 2000 *J. of Appl. Crystallography* **33**, 771-4
[49] Ghosh R E, and Rennie A R 1990 *Inst. of Physics Conf. Series* **107**, 233-44
[50] Lindner P 2002 *Neutron, X-rays and Light. Scattering Methods Applied to Soft Condensed Matter*, ed P Lindner and Th Zemb (North-Holland Delta Series, Elsevier 2002) pp 23

Journal of Physics: Conference Series **294** (2011) 012014 doi:10.1088/1742-6596/294/1/012014

Developments of In-Situ SEOP Polarized ^{3}He Neutron Spin Filter in Japan

H Kira[1 *], Y Sakaguchi[1,7], T Oku[1], J Suzuki[1], M Nakamura[1], M Arai[1], Y Endoh[1], L J Chang[1, 2, 3], K Kakurai[2], Y Arimoto[4], T Ino[4], H M Shimizu[4], T Kamiyama[4], K Ohoyama[5], H Hiraka[5], K Tsutsumi[5] and K Yamada[6]

[1]Japan Atomic Energy Agency, 2-4 Shirane, Shirakata, Tokai, Naka-gun, Ibaraki 319-1195, Japan

[2]Nuclear Science and Technology Development Center, National Tsing Hua University, Hsinchu 30013, Taiwan

[3]Department of Physics, National Cheng Kung University, Tainan 70101, Taiwan

[4]High Energy Accelerator Research Organization, 1-1 Oho, Tsukuba 305-0801, Japan

[5]Institute for Materials Research, Tohoku University, 2-1-1, Katahira, Aoba-ku, Sendai 980-8577, Japan

[6]WPI-Advanced Institute for Materials Science, Tohoku University, 2-1-1 Katahira, Aoba-ku, Sendai 980-8577, Japan

E-mail: kira.hiroshi@jaea.go.jp

Abstract. We launched the polarized ^{3}He neutron spin filters (NSF) project in order to provide neutron polarization for the pulsed neutron beams in Japan. We adopted the in-situ spin exchange optical pumping (SEOP) technique to polarize the nuclear spin of ^{3}He atoms because it has some advantages for our applications. The overall system size is compact and it avoids the problem of the time decay of nuclear spin of ^{3}He thus suppressing the costs of maintenance and providing other advantages [1, 2] with respect to data analysis and quality. In this paper, we performed pulsed neutron beam tests of our compact in-situ SEOP NSF system at the BL10 beamline in the Materials and Life Science Experimental Facility of the Japan Proton Accelerator Research Complex (J-PARC). The polarization of the ^{3}He gas reached was 73 % and a pump-up time constant of 9.5 h was observed. This paper is a status report about the development of in-situ SEOP NSF system for the pulsed beam at J-PARC.

1. Introduction

Polarized neutron scattering techniques are very important and powerful tools to study the structures of magnetic materials, soft matters and so on [3, 4, 5]. Because user operation of J-PARC started in December of 2008 it is urgently necessary for us to develop neutron spin filters for our pulsed neutron scattering experiments. To polarize the neutron beam, Heusler alloy crystals [6] and magnetic supermirror devices [7] have been used widely for several tens of years. But the Heusler alloy crystals only provide monochrome polarized neutron beams and the supermirror devices can not deal well with thermal (i.e. 1.8 Å mean wavelength) neutron beams. In the last several years, another method, polarized ^{3}He neutron spin filters (NSF), have been extensively developed and become popular [8]. NSFs work because ^{3}He has a very large absorption cross section for neutrons in the opposite spin state to that of the ^{3}He nucleus and largely transmit the spin-parallel state for reasonable ^{3}He polarizations while the neutron scattering cross section of ^{3}He is small. Compared with other spin filters such as

[7] Present address: Research Center for Neutron Science and Technology, Comprehensive Research Organization for Science and Society, Tokai, Ibaraki 319-1106, Japan

Heusler crystals and magnetic supermirrors, the ^{3}He gas NSF have the following advantages: it can polarize various energy of neutrons such as cold, thermal and epithermal neutrons and it can be used for wide area and large divergence neutron beams. These characteristics of polarized ^{3}He are convenient for use at J-PARC as a NSF. So we launched the project team aiming to provide polarized ^{3}He NSFs for the pulsed neutron beams in Japan at J-PARC.

There are two major techniques to polarize the ^{3}He nuclear spins, one is the spin exchange optical pumping (SEOP) technique [9] and the other is the metastability exchange optical pumping (MEOP) technique [10]. Comparing with these two techniques, the SEOP technique has some advantages as described below. The overall size of a SEOP system is small enough to install into the instruments and the initial investment cost is almost one tenth of that of a MEOP system. Further the system configuration and use is simple because it can polarize the NSF cell directly and thus does not need daily maintenance. Whereas if we where to prepare the NSF by using a MEOP system, we would need to compress and transfer the polarized ^{3}He gas into the NSF cells using a large system and then transfer the NSF to the neutron beam. In addition to these advantages, in-situ SEOP NSF system can avoid the time decay of the ^{3}He nuclear spin experienced when off line polarized gas is used. On the other hand, the present SEOP NSF system has some disadvantages. The productive capacity of the SEOP system is quite small and the SEOP system cannot polarize a complicated NSF cell such as a fan shape cell or cell with Si crystal windows, etc. whereas the MEOP based system is suitable for such cells, and can have a higher capacity per system. Despite these drawbacks, considering the advantages we adopted the in-situ SEOP technique to polarize the NSF cells.

2. Design of the in-situ SEOP system

The ^{3}He NSF will work as a neutron polarizer or a neutron spin analyzer. Most of the instruments in the J-PARC MLF typically have an incident neutron beam size of less than (40 x 40) mm^2. So in case of using the NSF as a polarizer for the injected beam, a cylindrical cell with a diameter of 50 mm is sufficient. In the case of polarisation analysis, because most of the instruments in the J-PARC MLF have large solid angle detector banks, the NSF cells are required to cover them. The shape of the NSF cells should become more complex such as a cylindrical cell with a large diameter, a fan shape cell, etc. It is thought that the development of the system that uses a small NSF cell is easier. So we decided to develop the in-situ SEOP NSF system for incident beam polarizer first.

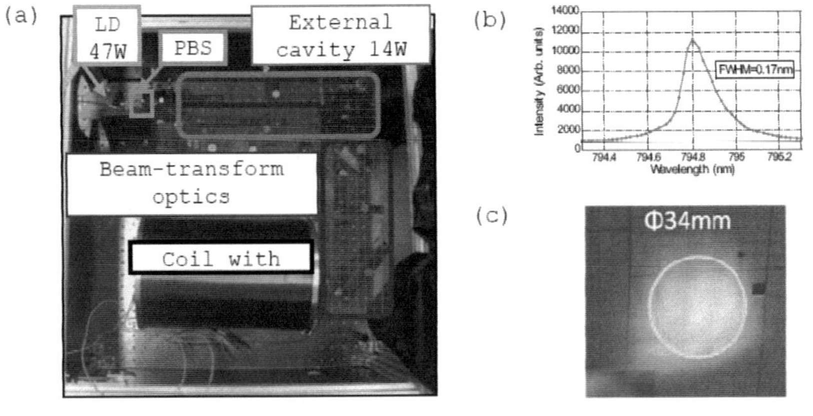

Figure 1. (a) Picture of the improved in-situ SEOP NSF system. (b) Wavelength and bandwidth of the laser was tuned by external cavity. (c) Picture of the laser spot at cell position.

Journal of Physics: Conference Series **294** (2011) 012014 doi:10.1088/1742-6596/294/1/012014

Fig. 1(a) shows the picture of the inside of the in-situ SEOP NSF system. The optics is put in the laser shield box made of black anodized aluminium board for laser safety. The laser shield box has a dimension of (60 x 60 x 30) cm^3. In this system we used a nLight VSA-100-796 diode laser (DL) with varied output power (up to 100 W at 120A) and a Newport laser driver model 5600 (up to 65 A). Because of the current limitation, the laser diode was driven at 47 W. The linear polarized laser beam emitted from the laser diode passes through a half-wave ($\lambda/2$) plate to rotate the direction of polarization and to change the ratio of p-wave to s-wave. Here we control the laser power of the p-wave component to 14 W. A polarizing beam splitter (PBS) separates the laser beam into the p-wave and the s-wave components [11]. Then the p-wave and s-wave components are sent to an external cavity and beam transform optics, respectively. The external cavity controls the wavelength and narrows the bandwidth of the laser. Fig. 1(b) shows the spectra of the emitted laser narrowed by the external cavity. The FWHM of the emitted laser was reduced to 0.17 nm. The beam-transform optics enlarge the laser beam spot size enough to illuminate the entire cell and converts the linearly polarized laser into circularly polarized laser light. Fig. 1(c) shows the laser beam spot at the cell position. We optimized the configuration of the optics for an NSF cell with a 34 mm diameter, which was used for the on-beam test. Details of the NSF cell are described next section. The beam spot size was sufficiently enlarged and the laser power density was more than 1 W/cm^2 over the entire cell. The cell is put into the oven in the solenoid coil, which can provide a magnetic environment with a transverse magnetic field gradient of less than 0.0005 cm^{-1} within a cylindrical space with a diameter of 50 mm and length of 100 mm for the ^{3}He gas NSF cell [2].

3. Polarization tests

We performed tests with a pulsed neutron beam on the NOBORU beamline at J-PARC MLF. Fig. 2 shows the schematic drawing and a picture of the experimental setup. In this test, we used a cylindrical NSF cell with a total ^{3}He gas quantity of 16.9 atm-cm. The dimensions of the NSF cell were 34 mm in diameter and 53 mm in length. The neutron beam monitor, NSF cell and resister divide photomultiplier tube (RPMT) 2-dimensional position sensitive detector (2D PSD) [12] were located 12.6 m, 13.0 m and 14 m far from the moderator, respectively. The beam size was set to (20 x 20) mm^2. The NSF cell was heated to 180 ºC during the experiment.

(a) (b) (c)

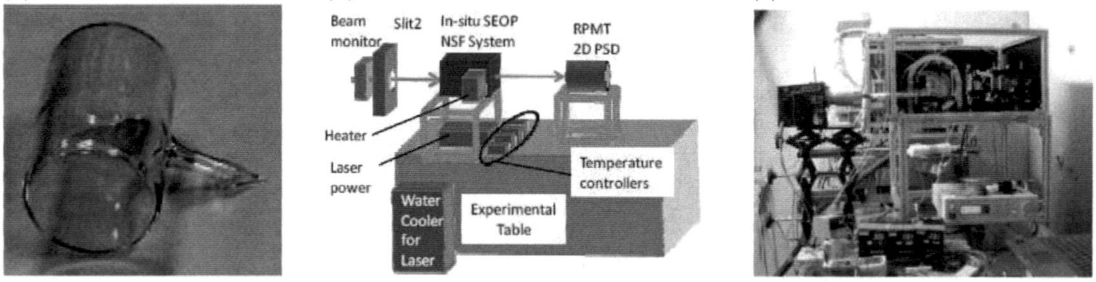

Figure 2. (a) Picture of the NSF cell with a diameter of 34 mm, length of 53 mm and total ^{3}He gas quantity of 16.9 atm-cm. (b) Schematic drawing of the experimental set up at BL10 beamline in J-PARC. (c) Picture of the in-situ SEOP system installed into BL10.

We performed optical pumping for 60 hours and then measured the wavelength dependence of the neutron beam transmission intensity through the NSF cell filled with polarized ^{3}He gas. After that we stopped the optical pumping, depolarized the ^{3}He gas in the NSF cell by using a permanent magnet and measured the neutron beam transmission intensity of the depolarized NSF cell. Results are shown in figure 3(a), the obtained data were normalized by the neutron beam monitor counts. The polarization of ^{3}He gas in the NSF cell is calculated from the neutron beam transmission intensity of the NSF cell filled with unpolarized ^{3}He gas T_0 and polarized ^{3}He gas T by following equations:

$$T = e^{-\sigma\rho t}\cosh P_{He}\sigma\rho t\,,$$
$$\frac{T}{T_0} = \cosh P_{He}\sigma\rho t \qquad (1)$$

where $T_0 = e^{-\sigma\rho t}$, P_{He} is the ^{3}He polarization, σ is the neutron absorption cross section for unpolarized ^{3}He, ρ is the number density of ^{3}He and t is the length of the ^{3}He gas [2]. The analyzed neutron beam polarization vs. neutron wavelength is shown in figure 3(b). The ^{3}He gas polarization was 73 % and independent of the wavelength, thus confirming the accuracy of the neutron calibrations. After adjusting the optics, we measured the pumping time dependence of the ^{3}He gas polarization for 50 hours. The result is shown in figure 3(c). The ^{3}He gas polarization finally reached to 70 % and a pump-up time constant of 9.5 h was observed. We performed a neutron beam test at the NOP beamline of the Japan Research Reactor-3 (JRR-3) research reactor in JAEA before. The NOP beamline can provide a highly polarized neutron beam (P>99 %) with a wavelength of 8.4 angstrom. We studied the dependence of ^{3}He polarization on temperature. Results of this study show that ^{3}He polarization was finally saturated at 73 % at an optimum temperature of 180 degree and the observed T_{UP} was 9.5 h. Those results are almost consistent with our previous experiments performed using a continuous neutron beam on the NOP port in the JRR-3 guide hall. The cause of the surpressed maximum polarization from 73 % to 70 % is considered to be due to improper adjustment of the optics.

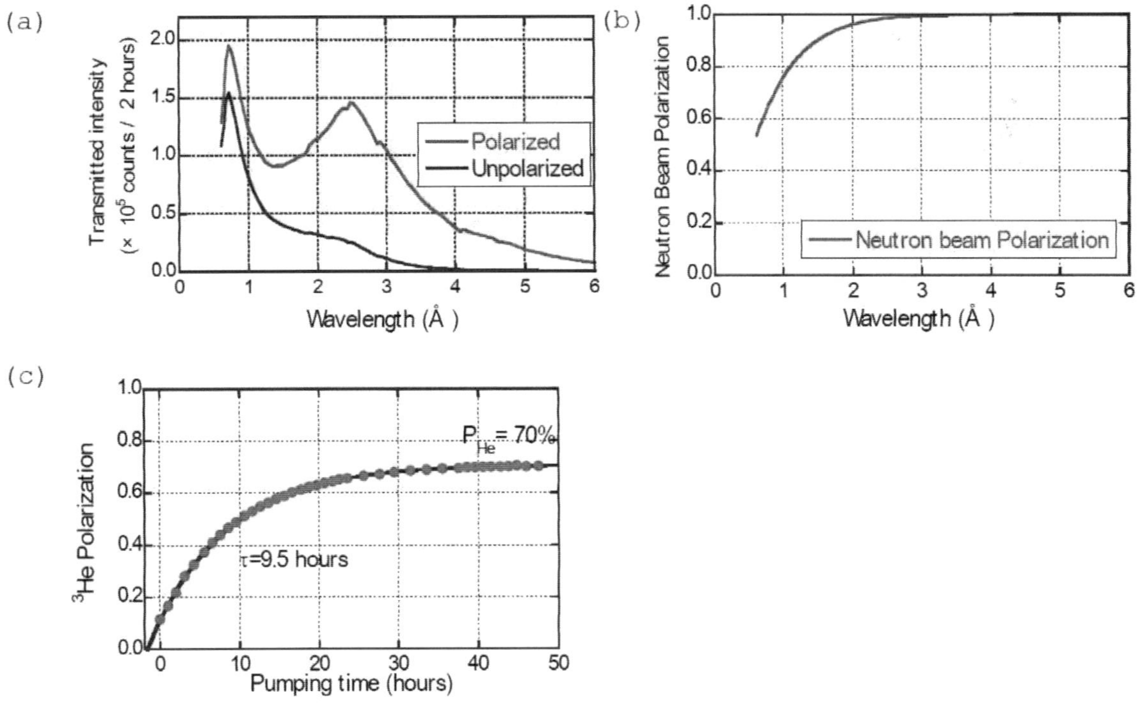

Figure 3. (a) Wavelength dependence of the transmitted neutron beam intensity for the NSF with polarized and depolarised ^{3}He gas. (b) Calculated neutron polarization. (c) Pumping time dependence of the ^{3}He gas polarization measured during in-situ SEOP.

4. Summary

We designed the compact in-situ SEOP NSF system aiming to use it as a polarizer for incident beams at J-PARC and performed a pulsed neutron beam tests of it at the J-PARC BL10 beamline. A ^{3}He gas polarization of 73 % and a pump-up time constant of 9.5 h was observed. These results were consistent with our previous tests performed at JRR-3 while using the continuous neutron beam there. In summary we succeeded to polarize the pulsed neutron beam at J-PARC by using the compact in-situ SEOP NSF system.

Acknowledgments

This work was supported by the Quantum Beam Fundamentals Development Program, MEXT.

References

[1] Babcock E, Mattauch S and Ioffe A 2011 *Nuclear Instruments and Methods A* **625** 43-6
[2] Ino T, Nakamura M, Oku T, Shinohara T, Suzuki J, Ohoyama K and Hiraka H 2009 *Physica B: Condensed Matter* **404** 2667
[3] Moon R M, Riste T and Koehler 1969 *Phys. Rev.* **181** 920
[4] Fong H F, Bourges P, Sidis Y, Regnault L P, Bossy J, Ivanov A, Milius D L, Aksay I A and Keimer B 2000 *Phys. Rev. B* **61** 14773-86
[5] Gabrys B J 1999 *Physica B: Condensed Matter* **267-268** 122
[6] Shull C G and Smart J S 1949 *Phys. Rev.* **76** 1256-7
[7] Sinclair R N and Brockhouse B N 1960 *Phys. Rev.* **120** 1638-40
[8] Cussen L D, Goossens D J and Hicks T J 2000 *Nuclear Instruments and Methods A* **440** 409-20
[9] Roberts T W, Humblot H, Tasset F, Lelievre-Berna E, Wildes A R, Ivanov A, Petoukhov A K, Regnault L P, Hillier AD, Kulda J, Stewart JR, Thomas M, Malbert P, Jullien D, Gibert Y, Gay R, Bourgeat-Lami E and Chung R 2001 *Physica B: Condensed Matter* **297** 282-7
[10] Walker T G and Happer W, 1997 *Rev. Mod. Phys.* **69**, 629-42
[11] Boag S, Babcock E, Andersen K H, Becker M, Charlton T R, Chen W C, Dalgliesh R M, Elmore S D, Frost C D, Gentile T R, Anton R L, Parnell S R, Petoukhov A K, Skoda M W A and Soldner T 2009 *Physica B: Condensed Matter* **404** 2659-62
[12] Hirota *K,* Shinohara T, Ikeda K, Mishima K, Adachi T, Morishima T, Satoh S, Oku T, Yamada S, Sasao H, Suzuki J and Shimizu H M 2005 *Physical Chemistry Chemical Physics* **7** 1836-8

Journal of Physics: Conference Series **294** (2011) 012015 doi:10.1088/1742-6596/294/1/012015

Polarized ^{3}He targets in medium energy physics at MAMI

J Krimmer[1,2*], W Heil[3], S Karpuk[3], E W Otten[3] and Z Salhi[3]

[1] Institut für Kernphysik, Johannes Gutenberg-Universität, Johann-Joachim-Becher-Weg 45, 55128 Mainz, Germany

[2] Institut de Physique Nucléaire de Lyon, 4 rue Enrico Fermi, 69622 Villeurbanne cedex, France

[3] Institut für Physik, Johannes Gutenberg-Universität, Staudinger Weg 7, 55128 Mainz, Germany

E-mail: j.krimmer@ipnl.in2p3.fr

Abstract. Experiments with polarized ^{3}He targets at the Mainz Microtron (MAMI) involve double polarized photoabsorption or the extraction of the electric form factor of the neutron $G_{e,n}$ via electron scattering, respectively. Polarized ^{3}He with an initial polarization of up to 70% at the experimental areas is provided by a polarizer based on the principle of metastability exchange optical pumping (MEOP). In this article both target setups for the photon- and the electron beamlines will be described and a status of the data analysis will be given.

1. Introduction

Polarized ^{3}He is used as an effective polarized neutron target in medium energy physics experiments, owing to the large S-state probability [1] where the two protons saturate their spins so that the spin of the neutron is aligned with the spin of the nucleus. This property can be used, e.g., in double polarized quasi elastic electron scattering to extract the electric form factor of the neutron $G_{e,n}$ via the reaction $^3\vec{\mathrm{He}}(\vec{\mathrm{e}},\mathrm{e}'\mathrm{n})\mathrm{pp}$.

Furthermore, experimental access to a validation of the Gerasimov-Drell-Hearn (GDH) sum rule[2; 3] of the neutron is possible via double polarized photoabsorption on polarized ^{3}He.

A polarizer based on the principle of metastability exchange optical pumping (MEOP), located in the Institute of Physics, provides polarized ^{3}He with a nuclear polarization of up to 76% at a flux of more than 1 bar·l/h [4; 5]. After filling at the polarizer, the target cells are brought to the experimental areas in an auxiliary magnetic holding field in Helmholtz configuration powered by a rechargeable battery. This remote type of operation requires long relaxation times of the target cells and a minimization of the polarization losses during the transport.

The experiments that will be described in the following sections have been performed at the MAMI accelerator in Mainz which provides electrons with a maximum energy of 1.6 GeV [6]. Electron scattering experiments (Section 2) with high resolution magnetic spectrometers are performed in the A1 hall, whereas a tagged photon beam in combination with a 4π-detector system is available in the A2 experimental area for the double polarized photoabsorption measurements (Section 3).

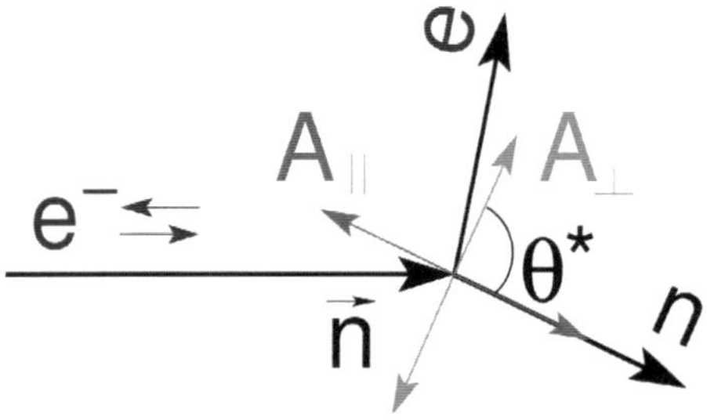

Figure 1. Double polarized electron scattering to extract $G_{e,n}$ via beam helicity asymmetries.

2. Polarized ^{3}He in the electron beam

2.1. Electron scattering

The differential cross section of electron scattering on extended spin-1/2 structures is described by the Rosenbluth formula:

$$\frac{d\sigma}{d\Omega} = \left(\frac{d\sigma}{d\Omega}\right)_{\text{Mott}} \left[\frac{G_e^2(Q^2) + \tau G_m^2(Q^2)}{1+\tau} + 2\tau G_m^2(Q^2)\tan^2\left(\frac{\theta}{2}\right)\right] \tag{1}$$

with $\left(\frac{d\sigma}{d\Omega}\right)_{\text{Mott}}$ the Mott differential cross section, Q^2 the (negative) transferred four-momentum squared and θ the scattering angle. $\tau = \frac{Q^2}{4M^2}$ is a kinematic factor. The so-called Sachs form factors G_e and G_m characterize the electromagnetic structure. In a special reference frame (the Breit frame) these are the Fourier transforms of the charge distribution and the magnetization density, respectively.

The angular dependence of Equation 1 can be used to separate G_e and G_m (Rosenbluth separation) via measurements of differential cross sections in various kinematical regimes. For an experimental access to the neutron electric form factor $G_{e,n}$ the Rosenbluth method is not applicable as the neutron is uncharged and therefore $G_{m,n}$ dominates over $G_{e,n}$ in Eq. 1. $G_{e,n}$ can be extracted, however, via beam helicity asymmetries. In Figure 1 the schematics of double polarized electron scattering is given. The scattered electron e' is detected in coincidence with the outgoing neutron n. The number of events $N^{\pm}$ divided by the luminosity $L^{\pm}$ is recorded for the two helicity states $(\pm)$ of the incoming electrons e. The beam helicity asymmetry A is then defined as

$$A = \frac{N^+/L^+ - N^-/L^-}{N^+/L^+ + N^-/L^-} \tag{2}$$

In plane wave impulse approximation (PWIA) this asymmetry can be related to the electromagnetic form factors [1]

$$A^{\text{PWIA}}(\theta^*, \phi^* = 0) = \frac{aG_{en}G_{mn}\sin(\theta^*) + bG_{mn}^2\cos(\theta^*)}{cG_{en}^2 + dG_{mn}^2} \tag{3}$$

where θ^* is the angle between the direction of the momentum transfer and the orientation of the nuclear spin of the target. a, b, c, d denote kinematical factors. In the parallel or perpendicular configurations, i.e. $\theta^* = 0°, 180°$ or $\theta^* = 90°, 270°$, respectively, Eq. 3 reduces to

$$A_{\parallel} = \pm\frac{bG_{mn}^2}{cG_{en}^2 + dG_{mn}^2} \qquad A_{\perp} = \pm\frac{aG_{en}G_{mn}}{cG_{en}^2 + dG_{mn}^2} \tag{4}$$

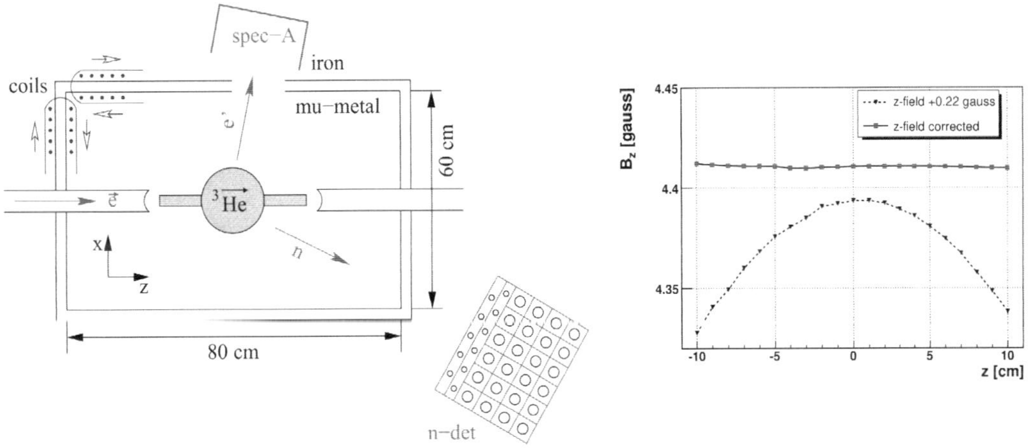

Figure 2. Left: target setup in the electron beamline, right: magnetic guiding field along the z-direction in the target cell region.

For the experimental asymmetries $A^{exp}_{\|,\perp}$ the beam and target polarizations P_e and P_n have to be taken into account. In the ratio

$$\frac{A_{\perp}{}^{exp}}{A_{\|}{}^{exp}} = \pm\frac{a}{b}\frac{G_{en}}{G_{mn}}\frac{\overline{P_eP_n}^{\perp}}{\overline{P_eP_n}^{\|}} \tag{5}$$

systematic uncertainties are reduced due to the occurrence of relative polarization values, only. $G_{m,n}$ has been measured with great precision in different experiments [7], so that $G_{e,n}$ can be obtained from Eq. 5 via a measurement of $A^{exp}_{\perp}/A^{exp}_{\|}$.

2.2. Setup

The magnetic guiding field for the polarized ^{3}He needs to meet the following requirements. The spin orientation needs to be rotated in the scattering plane for measurements of $A^{exp}_{\perp}$ and $A^{exp}_{\|}$. Furthermore, a sufficient homogeneity is required. When the relative field gradient $(dB/dz)/B_0$ is smaller than $5{\cdot}10^{-4}\text{cm}^{-1}$ the corresponding relaxation time due to field gradients T_1^{grad} is larger than 1000 hours for cells filled with 5 bar [8].

The target cells with the polarized gas are located in the middle of a box, made from μ-metal and iron plates with coils wound around, providing the magnetic guiding field [9] (see Fig. 2, left). The holes in the box for the primary electron beam and for outgoing particles spoil the perfect homogeneity of the magnetic field which would be obtained in a completely closed box in this configuration. By means of correction coils, a sufficient homogeneity can be regained, however, in the actual setup (see Fig. 2 right).

The target cells are manufactured from fused silica with a wall thickness of 2 mm. They comprise a spherical part with an outer diameter of 9 cm and two cylindrical sidearms resulting in a total length of 25 cm. The entry windows for the electron beam consist of 50 μm beryllium, covered by a thin layer (0.4 μm) of aluminum. Cesium coating [10] and a proper demagnetization procedure [11; 12] result in wall relaxation times T_1^{wall} between 100 and 200 hours.

For online polarization measurements two methods are available. A relative method, where the free induction decay (FID) signal is observed after tipping the $^3\vec{\text{He}}$ magnetization nonadiabatically by a small angle via a transverse field pulse. The second, absolute, method uses the measurement of the magnetic field generated by a dense sample of polarized gas [13]. In order to get rid off the magnetic holding field which is about a factor 1000 larger than the

Journal of Physics: Conference Series **294** (2011) 012015 doi:10.1088/1742-6596/294/1/012015

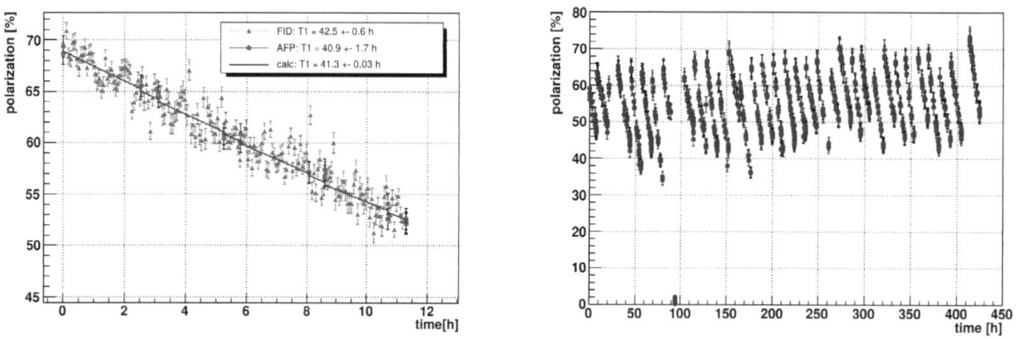

Figure 3. The polarization during a single turn (left) and during the complete beamtime (right) for the $G_{e,n}$ measurements.

field generated by the polarized gas, a 180 degree spin flip via adiabatic fast passage (AFP) is involved in this method. Both polarization measurement methods and the corresponding calibrations are described in Ref. [14] and are not repeated here.

2.3. Performance in the beam and status of data analysis
Besides the above mentioned relaxation due to field gradients and the interaction of the polarized gas with the container walls, under real experimental conditions also the relaxation due to the production of $^3He^+$ ions in a beam of charged particles [15] and the dipolar relaxation [16] has to be taken into account. For our setup in a 10 μA electron beam T_1^{beam} amounts to 220 hours[1] and T_1^{dipole} is 150 hours at a pressure of 5 bar. Adding up all contributions according to

$$\frac{1}{T_1^{total}} = \frac{1}{T_1^{grad}} + \frac{1}{T_1^{dipole}} + \frac{1}{T_1^{beam}} + \frac{1}{T_1^{wall}} \tag{6}$$

results in a total relaxation time T_1^{total} of 30-40 hours (see Fig. 3 left). The target cells have been exchanged twice per day, the temporal variation of the polarization during the complete beamtime for the $G_{e,n}$ measurements is displayed in the right part of Fig. 3. The initial polarization reached up to 72% and the mean polarization was about 55-60%. More details about the target setup for the electron beam can be found in Ref. [14].

The electric form factor of the neutron $G_{e,n}$ has been measured at $Q^2 = 1.58$ (GeV/c)^2. The data analysis which is part of a PhD thesis [17] is in progress. A preliminary data point is consistent with recent measurements [18] at higher Q^2 values.

3. Polarized ^{3}He in the photon beam

In the A2 experimental area a beam of energy tagged photons is available. The principle of tagging is displayed in the left part of Figure 4. The photons are produced via bremsstrahlung of electrons impinging on a thin radiator (e.g. 10μm Cu). The electrons are then deflected in a dipole magnet where the bending radius depends on the electron energy that is left after the bremsstrahlung process. Due to the hit position on a scintillator array in the focal plane of the magnet, the electron energy E_e is determined. With the known primary energy E_0 the energy of the photon E_γ is simply $E_\gamma = E_0 - E_e$ [19]. Circularly polarized photons are generated by bremsstrahlung of linearly polarized electrons on an amorphous radiator [20].

[1] The production of $^3He_2^+$ molecular ions has been suppressed by adding a small amount (10^{-2}) of N_2 as quenching gas.

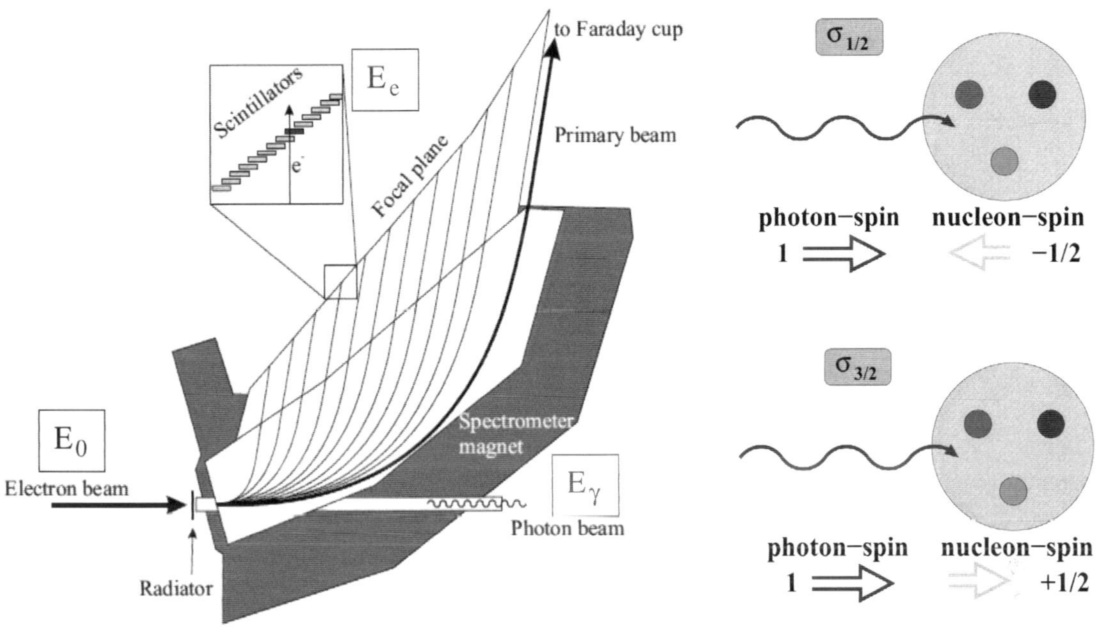

Figure 4. Left: production of energy tagged photons. Right: double polarized total photoabsorption.

3.1. GDH sum rule

One aim of the experimental program with the polarized ^{3}He target in the photon beam is the validation of the Gerasimov-Drell-Hearn (GDH) sum rule [2; 3] of the neutron. This sum rule has already been derived in the mid 1960s by using very fundamental physics principles, only. It reads

$$\int_{E_{\gamma,0}}^{\infty} dE_\gamma \frac{\sigma_{3/2} - \sigma_{1/2}}{E_\gamma} = \frac{2\pi^2\alpha}{m^2}\kappa^2 \tag{7}$$

with $\sigma_{3/2,1/2}$ the total photoabsorption cross sections of circularly polarized photons on longitudinally polarized nucleons (see Fig. 4, right side). The cross section difference is divided by the photon energy E_γ and integrated over the complete energy range. This means that the l.h.s. of Equation 7 includes the complete excitation spectrum of the nucleon, whereas on the r.h.s. only static properties like the mass m and the anomalous magnetic moment κ appear. α denotes the fine structure constant. An experimental check of the GDH sum rule for the proton has been performed at the accelerators MAMI in Mainz and ELSA in Bonn in the energy range from the threshold up to 3 GeV [21–23]. For an experimental access to the neutron the use of nuclear targets is inevitable. So far, only double polarized data from deuteron targets exist in a limited energy range [24; 25]. Here, the use of polarized ^{3}He provides an alternative, and due to the spin structure, more direct access to an experimental validation of the GDH sum rule for the neutron.

3.2. Setup

A sideview of the experimental setup at the photon beam is displayed in Figure 5. During data taking the cell with the polarized gas is located in the middle of the Crystal Ball detector [26].

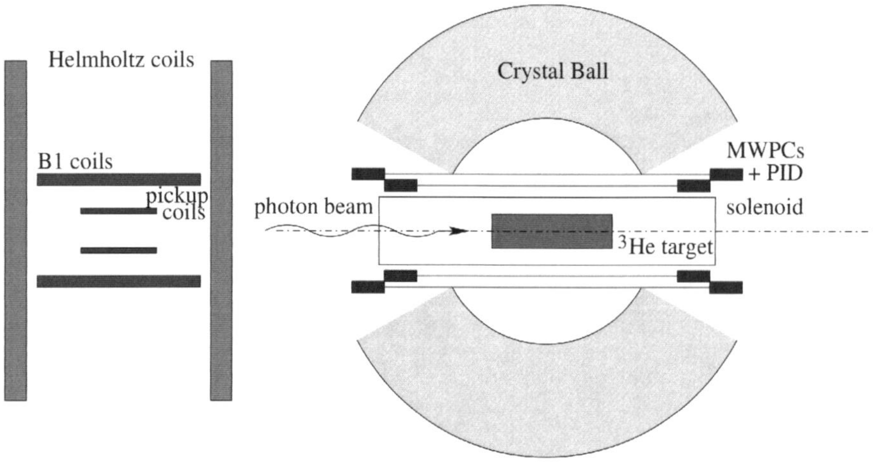

Figure 5. Experimental setup in the photon beam.

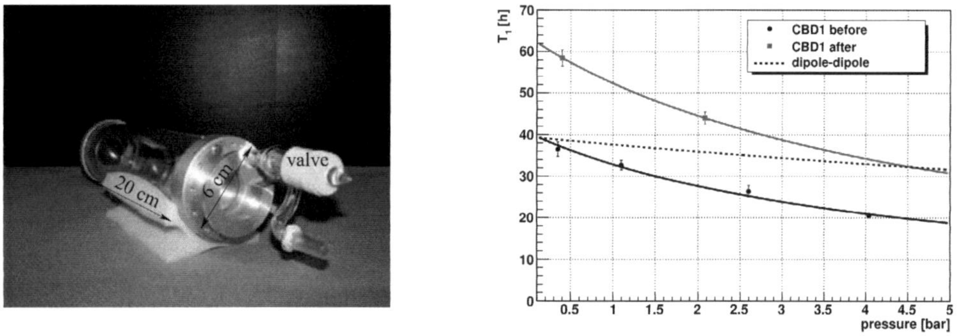

Figure 6. Left: Target cell for use in the photon beam. Right: Pressure dependence of T_1.

The magnetic guiding field is provided by a solenoid [27]. Due to lack of space inside the detector, polarimetry is performed outside the detector inside a pair of Helmholtz coils. An automatic transport system[2] has been installed for movement of the cell between the two positions. Inside the Helmholtz coils the decay of polarization is measured via the FID method (see Section 2.2). The complete polarization measurement cycle, including the movement of the cell, is completed within 3 minutes and causes only a 1 % relative polarization loss.

The target cells obey cylindrical geometry with a diameter of 6 cm and a total length of 20 cm (see Fig. 6 left side). They are made from quartz glass or pyrex and coated with cesium. Various materials for the entry windows for the photon beam have been tested. Besides tightness for the desired pressures and good properties regarding relaxation, the area density of the window foils should be as small as possible to suppress background events caused by interaction of the photon beam with the window material. 50 μm Mylar coated with aluminum seemed to be a good choice. It turned out, however, that these foils remain not tight after multiple exposures to high pressures. Consequently water vapor from the surrounding air can diffuse inside the cell and destroy the cesium coating. The high tensile strength metallic alloy Havar is compatible to Mylar regarding background events, though it is not ideal for keeping the ^{3}He polarized. The preferred

[2] Igus DryLin ZLW Zahnriemenachse Basic 02

solution is 50 μm titanium which resulted in the highest relaxation times so far. The measured T_1 times of cell 'CBD1' as a function of the pressure inside the cell is shown in the right part of Fig. 6. The T_1 times decrease stronger with increasing pressure than expected from dipole-dipole interaction [16] (dashed line). This effect can be attributed to ferromagnetic contaminants [28]. An improved demagnetization procedure led to an increase of the wall relaxation time for cell 'CBD1' from 40 to 60 hours (extrapolation of black and red lines in Fig. 6 to zero pressure).

3.3. Performance in the beam and status of data analysis

A first measurement with polarized ^{3}He in the photon beam has been performed in July 2009. It has been demonstrated that the complete setup works fine and that polarization losses during

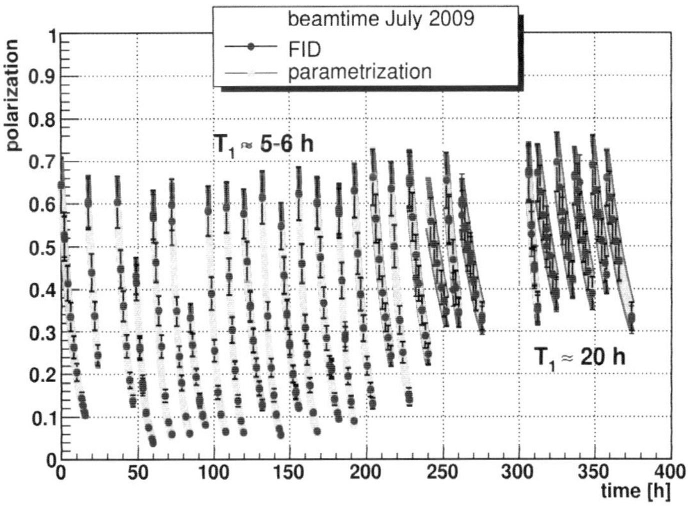

Figure 7. Polarization during the first beamtime in the photon beam in July 2009.

handling of the cells can be kept under control. The polarization performance during the complete beamtime is displayed in Figure 7. The polarization has been measured every two hours by means of the FID method. The agreement of the parametrization with the data points shows that there are no additional relaxation terms besides the known ones. The target cells have been exchanged twice per day. During the first part of the beam time only cells with total relaxation times between 5 and 6 hours were available, only, whereas in the last part cells with 20 hour livetime could be used. More details about the target setup for the photon beam will be given in an upcoming paper.

Data have been taken with primary electron energies of 855 and 525 MeV, respectively, covering the energy range of the delta resonance. The analysis of the data is part of a PhD thesis [27]. Preliminary results show a clear signal of the delta resonance.

4. Summary and outlook

In this article the setups for experiments with polarized ^{3}He in the electron and the photon beam at MAMI have been described. In the electron beam a measurement of $G_{e,n}$ at $Q^2 = 1.58(\mathrm{GeV/c})^2$ has been performed. For completeness it should be mentioned that also the reaction $^3\overrightarrow{\mathrm{He}}(\vec{\mathrm{e}},\mathrm{e}'\vec{\mathrm{p}})\mathrm{d}$ has been studied [29] with a similar setup.

In the photon beam the first experiment with a polarized ^{3}He target in the energy range of the delta resonance has been performed which demonstrated the functionality of the assembly.

For future experiments the improvement of the mean polarization is worthwhile. A major improvement can be made by installing the new compact polarizer [30] directly at the experiment. This *online* polarization would allow for quasi constant polarization values around 70 %.

Acknowledgments
This work was supported by the Deutsche Forschungsgemeinschaft under Contract SFB 443.

References
[1] Blankleider B and Woloshyn R M 1984 *Phys. Rev.* **C 29** 538

[2] Gerasimov S 1966 *Sov. J. Nucl. Phys.* **2** 430

[3] Drell S D and Hearn A C 1966 *Phys. Rev. Lett.* **16** 908–911

[4] Batz M, Baeßler S, Heil W, Otten E W, Rudersdorf D, Schmiedeskamp J, Sobolev Y and Wolf M 2005 *J. Res. Natl. Inst. Stand. Technol.* **110** 293

[5] Wolf M 2004 Ph.D. thesis University of Mainz

[6] Jankowiak A 2006 *Eur. Phys. J.* **A 28**(s01) 149

[7] Lachniet J *et al.* (CLAS Collaboration) 2009 *Phys. Rev. Lett.* **102** 192001

[8] Schearer L D and Walters G K 1965 *Phys. Rev.* **A 139** 1398

[9] Hiebel S, Großmann T, Kiselev D, Schmiedeskamp J, Gusev Y, Heil W, Karpuk S, Krimmer J, Otten E W and Salhi Z 2010 *Journal of Magnetic Resonance* **204** 37 – 49

[10] Deninger A, Heil W, Otten E W, Wolf M, Kremer R K and Simon A 2006 *Eur. Phys. J.* **D 38** 439

[11] Schmiedeskamp J, Elmers H J, Heil W, Otten E W, Sobolev Y, Kilian W, Rinneberg H, Sander-Thmmes T, Seifert F and Zimmer J 2006 *Eur. Phys. J.* **D 38** 445

[12] Thiel F, Schnabel A, Knappe-Gruenberg S, Stollfuss D and Burghoff M 2007 *Rev. Sci. Instrum.* **78** 035106

[13] Wilms E, Ebert M, Heil W and Surkau R 1996 *Nucl. Instr. Methods* **A 401** 491

[14] Krimmer J, Distler M, Heil W, Karpuk S, Kiselev D, Salhi Z and Otten E W 2009 *Nucl. Instr. Methods* **A 611** 18–24

[15] Chupp T, Loveman R, Thomson A, Bernstein A and Tieger D 1992 *Phys. Rev.* **C 45** 915–932

[16] Newbury N R, Barton A S, Cates G D, Happer W and Middleton H 1993 *Phys. Rev.* **A 48** 4411

[17] Schlimme S in preparation Ph.D. thesis University of Mainz

[18] Riordan S *et al.* 2010 (*Preprint* `nucl-ex/1008.1738`)

[19] McGeorge J C *et al.* 2008 *Eur. Phys. J. A* **37** 129–137

[20] Olsen H and Maximon L C 1959 *Phys. Rev.* **114** 887–904

[21] Ahrens J *et al.* 2001 *Phys. Rev. Lett.* **87** 022003

[22] Dutz H *et al.* 2003 *Phys. Rev. Lett.* **91** 192001

[23] Dutz H *et al.* 2004 *Phys. Rev. Lett.* **93** 032003

[24] Ahrens J *et al.* 2006 *Phys. Rev. Lett.* **97** 202303

[25] Dutz H *et al.* 2005 *Phys. Rev. Lett.* **94** 162001

[26] Starostin A *et al.* 2001 *Phys. Rev. C* **64** 055205

[27] Aguar-Bartolome P 2011 Ph.D. thesis University of Mainz

[28] Schmiedeskamp J 2005 Ph.D. thesis University of Mainz
[29] Weinriefer M in preparation Ph.D. thesis University of Mainz
[30] Mrozik C, Edner O, Hauke C, Heil W, Karpuk S, Klemmer J and Otten E W 2011 *IOP journal of physics conference series, this volume*

Tunable magnetostatic cavity for housing a ^{3}He neutron spin filter

S Masalovich[1*] **and O Lykhvar**[1,2]
[1]Technische Universität München, Forschungsneutronenquelle Heinz Maier-Leibnitz (FRM II), Lichtenbergstr. 1, 85747 Garching, Germany

E-mail: Sergey.Masalovich@frm2.tum.de

Abstract. A new design of a magnetostatic cavity for housing a neutron spin filter with spin-polarized ^{3}He gas at a neutron instrument is proposed. Movable components made of soft magnetic materials allow for an accurate tuning of the homogeneity of the holding magnetic field within the cavity and, thus, ensure a long ^{3}He spin relaxation time. In addition, this design enables to correct the holding field in the presence of an external magnetic source near the cavity, which is usually the case in neutron experiments with polarization analysis.

1. Introduction

Recent years have seen significant progress in the application of neutron spin filters (NSF) utilizing polarized ^{3}He gas for neutron polarization analysis [1-10]. The straight-line passage of a neutron beam through a NSF with no change of neutron trajectory enables one to measure a neutron polarization for nearly any divergent scattered beam. Furthermore, the NSFs offer homogeneous analyzing efficiency with a predictable value, predictable transmission, a negligible small angle scattering from the ^{3}He cell and produce only a low gamma-ray background. Taken together these features make the NSF technique extremely useful in applications such as neutron imaging with polarization analysis, small-angle neutron scattering, off-specular reflectometry and large solid angle polarization analysis.

The successful application of ^{3}He NSFs in neutron experiments is based on the considerable improvements in the large-scale production of hyperpolarized ^{3}He gas in both Metastability Exchange Optical Pumping (MEOP) and Spin Exchange Optical Pumping (SEOP) methods. In the MEOP method, direct optical pumping of metastable ^{3}He atoms occurs in a weakly ionized ^{3}He plasma at about 1 mbar gas pressure [11-13] and hence a subsequent compression is required to produce a dense polarized gas. The gas at a pressure of a few bars may be then collected in a detachable cell (glass container) of a given size and shape and transported to a neutron instrument. The high production rate allows for several neutron experiments to be served in parallel. The limitation of this approach, however, is the necessity of the cell replacement in the course of a long running experiment since the polarization of the ^{3}He gas in the cell decays exponentially with time according to the equation

$$P(t) = P(0) \cdot \exp\left(-\frac{t}{T_1}\right) \tag{1}$$

---

[2] Present address: TOPTICA Photonics AG, 82166 Graefelfing / Munich, Germany

Here $P(0)$ – the initial ^{3}He gas polarization and T_1 – the longitudinal spin-relaxation time which will be discussed later.

Unlike MEOP, in the SEOP method ^{3}He gas can be polarized directly at high pressure and there is no need for an additional compression. Here, a production of a highly polarized ^{3}He gas is realized through the optical pumping of alkali-atoms (pure Rb, pure K or Rb-K mixture) in a high-pressure alkali-^{3}He gas mixture [14-17]. The polarized alkali-atoms then transfer polarization to the ^{3}He-nuclei in a spin-exchange process thus obtaining a dense spin-polarized ^{3}He gas. Although the polarization build-up time in SEOP is longer compared to MEOP, the SEOP method is of particular interest in neutron instrumentation since it allows for a ^{3}He gas to be polarized steadily on a neutron beam in the course of a running experiment and, thus, there is no requirement for cell replacement. This means, on the other hand, that the SEOP setup has a limited capability to serve several experiments in parallel.

In spite of a difference between these methods, there is one common requirement which is valid in both cases: a cell with ^{3}He gas has to be housed in a very homogeneous holding magnetic field during both the optical pumping and the neutron experiment. In the case of a detachable cell (e.g., MEOP cell), the inhomogeneity of the magnetic field sets an upper limit for the relaxation time of the gas polarization in the cell and, hence, can lead to either the necessity of a very frequent replacement of the cell or even to unacceptable conditions to perform the experiment. In the case of the on-line gas polarization (e.g., SEOP), the shortening of the relaxation time due to the inhomogeneity of the magnetic field results in lowering of the maximal achievable gas polarization in the cell. Therefore, the homogeneity of the holding magnetic field is of critical importance in production and implementation of the polarized ^{3}He gas. With the aim to provide such a homogeneous field, different manetostatic cavities were developed.

2. Magnetostatic cavity

A magnetostatic cavity is an important unit intended to produce a homogeneous magnetic field for holding a polarized ^{3}He NSF and to screen the polarized gas against environmental stray magnetic fields at a neutron instrument. Generally, the relaxation rate of the gas polarization in a NSF cell may be represented as a sum (see, for example, [18]):

$$\frac{1}{T_1} = \frac{1}{T_{cell}} + \frac{1}{T_{mag}} \tag{2}$$

Here T_{cell} – the relaxation time related to the cell (relaxation due to interactions of the gas atoms with the cell walls [19-23] and due to dipole-dipole interactions of helium atoms [24]) and T_{mag} – the relaxation time related to the gradient of the magnetic field over the cell volume [25-27]. The last term in equation (2) can be written as [18]

$$\frac{1}{T_{mag}[hour]} = 1.7\cdot 10^4 \left(\frac{|\nabla B_\perp|}{B_0} [cm^{-1}] \right)^2 \cdot \frac{1}{p[bar]} \tag{3}$$

where p is the gas pressure, B_0 refers to the value of the holding magnetic field and $B_\perp$ - to the component of the magnetic field normal to B_0. From hereon the ratio $|\nabla B_\perp|/B_0$ is referred to as a field gradient. It follows from equation (3) that a field gradient less than $3\cdot 10^{-4}$ cm^{-1} is required to ensure $T_{mag} > 500$ h in a 1 bar pressure cell.

Magnetostatic cavities aimed at providing a homogeneous magnetic field have been developed in all groups working with polarized gas (see, for example, [15, 28-33]). Generally, if we consider only non-cryogenic cavities, they can be classified as current based or permanent magnet based cavities depending on the source chosen to produce a holding magnetic field. Soft magnetic materials such as iron or mu-metal are used to redistribute the field over the desired volume and/or to screen against environmental fields. Typically, all these systems are based on an accurate design of a cavity with a fixed geometry and with an assumption made about the value and space distribution of the environmental magnetic field at a neutron instrument. This final assumption requires some free parameters to be added to the design. In many instances, however, these parameters cannot be evaluated in advance. As a result, the real field gradients over the cell volume may differ significantly at different instruments.

The simple method to solve this problem would be to built a magnetostatic cavity with an excessive screening factor and, hence, to make it heavy and/or large. One may expect this method to be extremely useful for a fixed cavity integrated to a neutron instrument or for a portable cavity designed to hold small cells (of size of a few centimeters). However, as it was mentioned earlier, the special benefit of using a polarized ^{3}He NSF is expected in such areas as a neutron imaging with polarization analysis, small-angle neutron scattering, off-specular reflectometry and large solid angle polarization analysis. In all these cases the use of large area NSF cells would be advantageous. Obviously, for relatively large cells the restrictions on the size and weight of the portable cavity make the method of excessive screening of limited usefulness.

In this paper we present a new approach to the design of a compact magnetostatic cavity which allows one to build a portable magnetostatic cavity for relatively large NSF cells.

3. Tunable magnetostatic cavity

With the aim of simplifying the use of NSFs at neutron instruments, a robust compact magnetostatic cavity was developed at the FRM II (Garching, Germany) (figure1). It is comprised of plain steel plates assembled in a box as shown in figure 2. Two rows of permanent magnets are sandwiched between steel plates on the sides. Each row consists of many pieces of permanent magnets assembled along the direction normal to the plane of the picture. The polarities of the magnets in both rows are identical and aligned in the vertical direction. The magnetic field is guided to the top and bottom parallel steel plates (5 and 6 in figure 2) and from there through the empty cavity, creating a holding field inside the box. In addition to this desired field, a strong stray field occurs in the vicinity of the permanent magnets resulting in an inhomogeneous field distribution. This is illustrated in figure 3 where the magnetic flux lines distribution was calculated for a cavity without inner mu-metal plates.

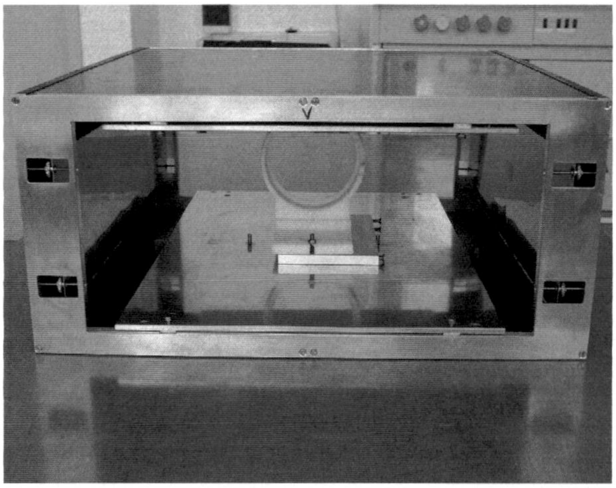

Figure1. TUnable Magnetostatic cavity (TUM box) at the FRM II. A neutron spin filter holder is located in the centre of the box.

Journal of Physics: Conference Series **294** (2011) 012016 doi:10.1088/1742-6596/294/1/012016

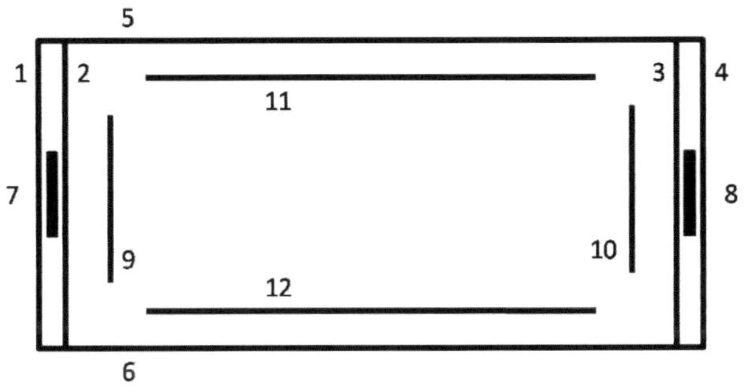

Figure 2. Front view of the TUM box. 1-6 – the plain steel plates, 7 and 8 – two rows of permanent magnets, 9-12 – the adjustable mu-metal plates.

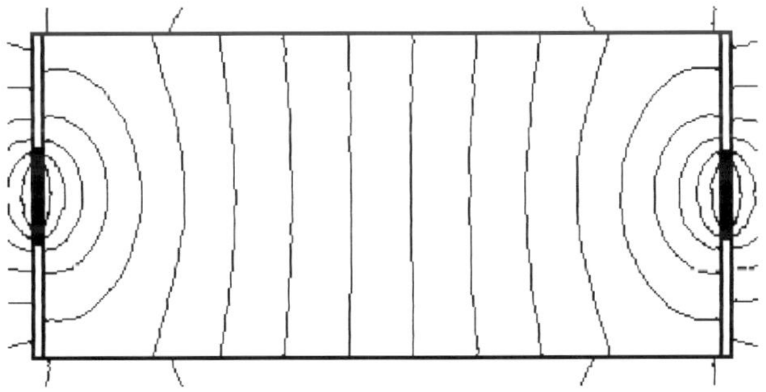

Figure 3. Magnetic field flux lines distributed over the box without mu-metal plates.

To correct the field distribution and build up a uniform field inside the box, four mu-metal plates (9-12) were installed as shown in figure 2. The thickness of these plates and the distance to the walls was chosen to affect the stray field in such a way that the resulting field inside the box becomes uniform. Figure 4 shows the distribution of the field inside the box calculated for the adjusted distances between the mu-metal plates and the walls. It can clearly be seen from this figure that some of the magnetic flux lines are closed by the mu-metal plates 9 and 10 and do not penetrate into the interior part of the cavity; this depends on both the thickness of the plates and the distances to the magnets. Optimizing the positioning of the mu-metal plates results in a very homogeneous field in the central area where a NSF cell is supposed to be located. The small shift of the mu-metal plates 9 and 10 in the direction of the sidewalls or in the opposite direction leads to the variation of the number of the closed lines and to deterioration of the field homogeneity.

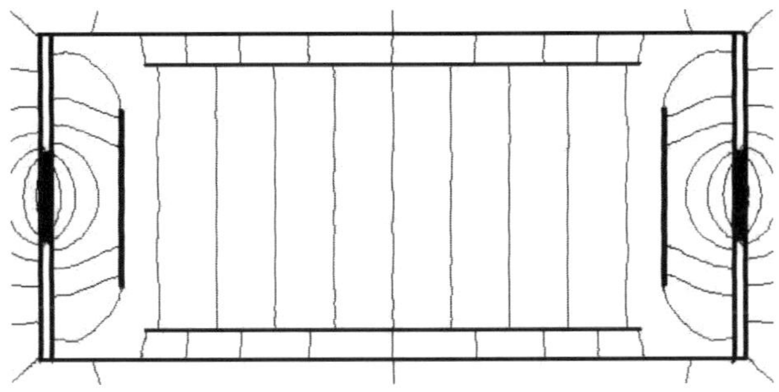

Figure 4. Magnetic field flux lines distributed over the TUM box with mu-metal plates installed and adjusted.

So far we have considered a model where all components have been thought to be ideal, i.e., all magnets had the same strength and perfectly aligned and all soft magnetic materials (plain steel and mu-metal plates) had a uniform permeability to magnetic field. In practice, however, one usually has to deal with non-ideal components like non-uniform permanent magnets and soft magnetic materials. Therefore, the practical design of a magnetostatic cavity has to be adapted to these conditions. The cavity in figure 2 was designed to meet this requirement. For this purpose, the positions and inclinations of the mu-metal plates 9 and 10 with respect to the sidewalls can be adjusted in situ to minimize the field inhomogeneity caused by the imperfection of the materials. This allows for a fine tuning of the magnetic field homogeneity inside the box, which may be, thus, referred to as a TUnable Magnetostatic box (TUM box).

Two additional mu-metal plates (11 and 12 in figure 2) are positioned near the bottom and the top of the box. Figure 5 illustrates the effect of these plates on the distribution of the holding field B_0 measured along the central line of the constructed box (along the neutron beam).

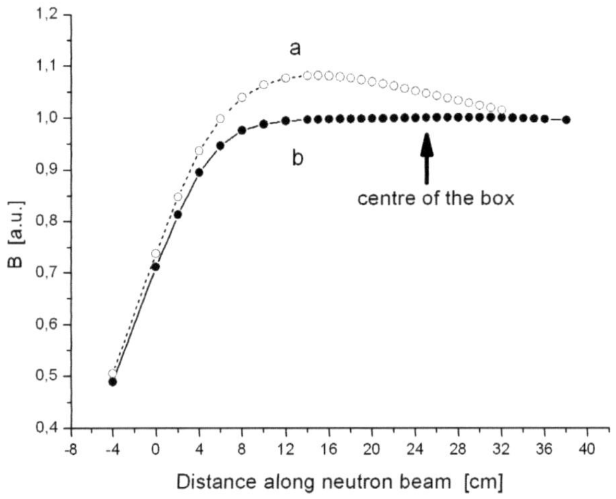

Figure 5. Effect of the horizontal mu-metal plates on the field homogeneity. Line "a" shows the field distribution measured without horizontal mu-metal plates and line "b" – after the horizontal plates were installed.

In this particular example the distribution of the vertical field component was measured first without mu-metal plates 11 and 12 (line "a") and then with mu-metal plates installed (line "b"). As one can see, a significant field gradient occurs in the central area of the box without mu-metal plates. This most probably results from the non-uniformity of the permeability of the steel plates. The mounting of the horizontal mu-metal plates helps to redistribute the field leading to a decrease of the field gradient. The adjustment of these plates in situ provides a way of further decreasing the field gradient to an acceptable value.

The first tunable magnetostatic box at the FRM II had dimensions 50x50x22 cm (WxLxH) and was capable of holding a cell with a gas volume of size 10x10x10 cm. The optimal design of the box was found with the help of the computer simulation software FEMM (Finite Element Method Magnetics [34]). For this optimal design the magnetic field gradient was calculated to be around $1 \cdot 10^{-4}$ cm^{-1}. In practice, however, the cavity assembled in accordance with this design displayed extremely poor magnetic field homogeneity over the desirable cell volume. We ascribe this largely to the inhomogeneous permeability of the plain steel plates and, to a smaller degree, to the variation of the magnetization of the pieces of permanent magnets since the latter were selected and distributed evenly in both rows of magnets. To reduce the field inhomogeneity (or, in other words, to tune the cavity), both the vertical and horizontal mu-metal plates were shifted and tilted against the calculated positions. This correction was performed by an iterative process with measuring the field gradient distribution after each step. Finally, for this particular cavity, the tuning was performed successfully with the resulting field gradient $4.4{\cdot}10^{-4}$ cm^{-1}. (Later we have constructed two more cavities with field gradients of $2.5{\cdot}10^{-4}$ cm^{-1} over the volume 10x10x10 cm^3.) It should be mentioned that the gradients of the field components normal to the main component B_0 (see equation (3)) were measured only along the sides of the cubic volume 10x10x10 cm in the central area of the box. This is a useful approach to perform express analysis of the field homogeneity and it offers fast tuning if necessary. This approach is based on the fact that the points located inside the volume are at greater distances from the sources of the magnetic field and hence we may expect the gradients at these points to be less than the ones measured on the outer surface of the volume. Thus, the mean value of the field gradient measured along the sides can be considered as an upper limit of the mean value of the field gradient over the entire cubic volume. Indeed, for the considered cavity, the field gradient in a smaller volume (6x6cm cross-section) was measured to be $3.5{\cdot}10^{-4}$ cm^{-1}.

The final test of the performance of the TUM box was done by measuring the relaxation time of polarized ^{3}He gas in a cell which was placed in the central area of the box. The polarization of the gas was performed at HELIOS, the facility for large-scale production of dense highly polarized ^{3}He gas at the FRM II (Garching, Germany) [7]. With the aim to measure gas polarization in the cell a very small amount of the gas was taken from the cell back to the HELIOS facility. The polarization was then evaluated by measuring the degree of circular polarization of the 668nm spectral line [35-36] emitted by the polarized gas in a low-pressure discharge at HELIOS. For the purpose of measuring the relaxation time, the polarization of the gas in the cell was measured successively over the period of two-three days. Fitting the experimental data to the exponential decay function (equation (1)) yields the relaxation time.

As mentioned above, the measured relaxation rate (inverse value of a relaxation time) represents the total effect of many parameters (see equation (2)). With the aim to evaluate the relaxation time related solely to the magnetic field gradient in the TUM box, we measured first the relaxation time in the cell positioned in the box and then the relaxation time in the same cell positioned in a very homogenous magnetic field at HELIOS. This second measurement was performed without repolarization of the gas. Since the magnetic field gradient at HELIOS is very low (about $1{\cdot}10^{-4}$ cm^{-1}), it results in a negligible contribution to the total relaxation rate. Thus, we may assume the relaxation rate measured at HELIOS is entirely related to the first term in equation (2). Thus, measuring T_1 in the TUM box and then T_{cell} in the magnetic field at HELIOS one can extract T_{mag}, i.e., relaxation time related to the field gradient in the TUM box. Applying equation (3) to the evaluated value of T_{mag}, the field gradient over the cell volume can be calculated. The following results were obtained: $|\nabla B_\perp|/B_0 \cong 5{\cdot}10^{-4}$ cm^{-1} over a cell with 10x11 cm cross-section and 5.5 cm length and $|\nabla B_\perp|/B_0 \cong 3.5{\cdot}10^{-4}$ cm^{-1} over a cell with 6 cm diameter and 10 cm length. These values are in a good agreement with the direct measurement of the field gradients discussed earlier. Thus, the measurements of the relaxation time confirmed the ability to tune the magnetic field homogeneity in the cavity, which showed initially a very poor performance. It is worth noting that after all these measurements have been finished we performed further adjustments of the field that brought the gradient down to the value $3.5{\cdot}10^{-4}$ cm^{-1} over the volume 10x10x10 cm^3.

4. Tuning in the presence of an external magnetic field

The approach with a tunable magnetic field can be readily applied to the magnetostatic cavity located in the vicinity of an external magnetic source, which is usually the case at a neutron instrument with polarization analysis. This external field usually originates from a neutron spin guiding magnetic system and/or from a magnetic field applied to the sample under study. The external magnetic field may result in an additional field gradient over the NSF cell in the cavity and special care has to be taken to minimize this effect. We have mentioned earlier the problems that may arise while constructing a passive shielding for cells of a relatively large size. In this respect the tunable magnetostatic cavity offers an alternative solution, since it was specially designed to correct any inhomogeneity of the holding field regardless of the particular source of that inhomogeneity. Consequently, the field gradient resulting from the presence of an external magnetic field can also be corrected in this cavity. With the aim to check this possibility, the TUM box was exposed to an external magnetic field similar to the field used in neutron experiments. In figure 6 the local source of the field is located at a distance D from the TUM box. This source represents a model of a guiding field with the size 20 x 32 cm (W x H) and the strength of 28 G.

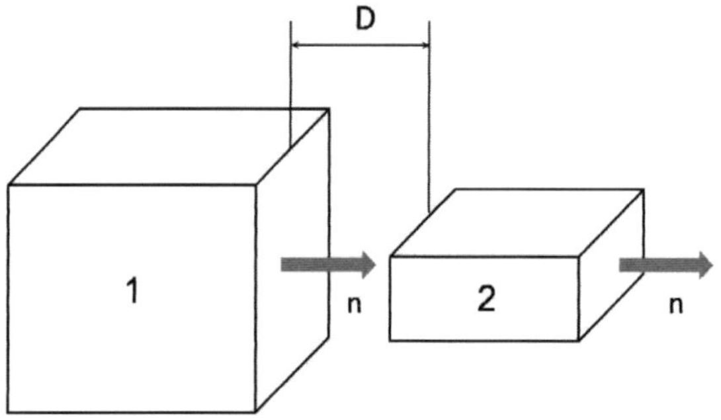

Figure 6. Experimental setup for measuring the effect of the external source. 1 – external source (magnetic guiding field at a neutron instrument), 2 – TUM box with a NSF cell. Neutron beam is shown for the reference only.

It is anticipated that the passive suppression of the external field within the TUM box can be performed only to some extent and the field gradient is expected to rise as an external source moves closer. Figure 7 presents the measured field gradient in the central area of the TUM box as a function of the distance D.

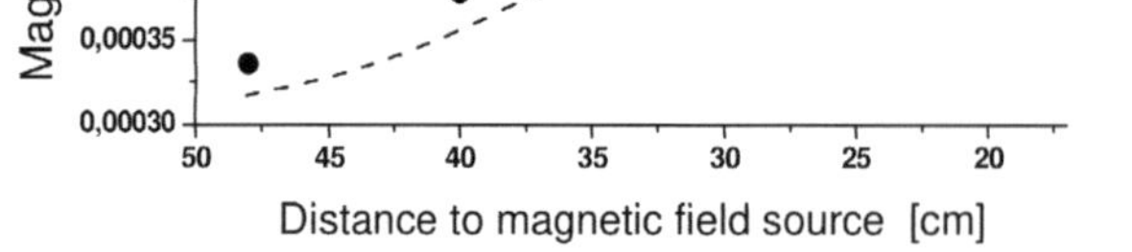

Figure 7. Effect of tuning in the presence of an external magnetic source. Bold circles – the measured gradient of the holding field averaged over the cell volume (10x10x10 cm). The dashed curve is a guide to the eye.

It is clearly seen that the field gradient over the cell volume increases as the distance between the TUM box and the external source decreases. At the distance of 22 cm the measured field gradient is twice as much as that without external source. The value of the magnetic field produced by the external source at the front of the box is 5 G. After the gradient has been doubled, the additional tuning with the help of the movable mu-metal plates in the TUM box was performed and the gradient was brought down to the required value (see the bold arrow in figure 7). This additional tuning was done with a simple tilting of the horizontal mu-metal plates at the top and bottom of the box. Finally, the obtained result proves the ability to tune a holding field at a neutron instrument.

5. Conclusion

The new design of a magnetostatic cavity with a homogeneous magnetic field is proposed and tested. The distinguishing feature of this design is the possibility for correcting the field gradient within the central area of the cavity in situ. This correction is performed with the help of special movable components, which are made of soft magnetic materials and integrated into the cavity. So, the design involves both passive and tunable shielding components. This fact is of special importance in the case when the cavity is exposed to an external magnetic field, which usually occurs at a neutron instrument with polarization analysis. The correction of the field gradient caused by an external source has been demonstrated in a special measurement performed with field geometry similar to one at a neutron instrument.

One may expect that replacing of the plain steel plates by mu-metal plates will result in even better performance of the tunable magnetic boxes and allows for more compact design for relatively large NSF cells.

Acknowledgments

This research project was supported by the European Commission under the 6th Framework Programme through the Key Action: Strengthening the European Research Area, Research Infrastructures. Contract no: RII3-CT-2003-505925.

The authors appreciate the critical review of the manuscript by Dr. Richard Mole.

References

[1] Rich D R, Fan S, Gentile T R, Hussey D, Jones G L, Neft B, Snow W M and Thompson A K 2001 *Physica* B*: Condensed Matter* **305** 203-7

[2] Heil W, Andersen K H, Cywinski R, Humblot H, Ritter C, Roberts T W and Stewart J R 2002 *Nucl. Instr. Meth.* A **485** 551-70

[3] Genile T R, Babcock E, Borchers J A, Chen W C et al 2005 *Physica* B*: Condensed Matter* **356** 96-102

[4] Masuda Y, Ino T, Skoy V R and Jones G L 2005 *Appl. Phys. Lett.* **87** 053506

[5] Andersen K H, Cubitt R, Humblot H, Jullien D, Petoukhov A, Tasset F, Schanzer C, Shah V R and Wildes A R 2006 *Physica* B*: Condensed Matter* **385-386** 1134-7

[6] Chupp T E, Coulter K P, Kandes M et al. 2007 *Nucl. Instr. Meth.* A **574** 500-9

[7] Hutanu V, Masalovich S, Meven M, Lyhkvar O, Borchert G and Heger G 2007 *Neutron News* **18** 14-16

[8] Keiderling U, Wiedenmann A, Rupp A, Klenke J and Heil W 2008 *Meas. Sci. Technol.* **19** 034009

[9] Gaspar A M, Busch S, Appavou M-S, Haeussler W, Georgii R, Su Y and Doster W 2010 *Biochimica et Biophysica Acta* **1804** 76-82

[10] Schulz M, Neubauer A, Masalovich S, Mühlbauer M, Calzada E, Schillinger B, Pfleiderer Ch and Böni P *J. Phys.: Conf. Ser.* **211** 012025

[11] Colegrove F D, Schearer L D and Walters G K 1963 *Phys.Rev.* **132** 2561-72

[12] Roberts T W, Humblot H, Tasset F, Lelievre-Berna E et al. 2001 *Physica* B*:*

Condensed Matter **297** 282-7
[13] Batz M, Baeßler S, Heil W, Otten E W, Rudersdorf D, Schmiedeskamp J, Sobolev Yu and Wolf M 2005 *J. Res. Natl. Inst. Stand. Technol.* **110** 293-8
[14] Bouchiat M A, Carver T R and Varnum C M 1960 *Phys. Rev. Lett.* **5** 373-5
[15] Chen W C, Gentile T R, O`Donovan K V, Borchers J A and Majkrzak C F 2004 *Review of Scientific Instruments* **75** 3256-63
[16] Jones G L, Dias F, Collet B, ChenW C, Gentile T R et al. 2006 *Physica* B*: Condensed Matter* **385-386** 1131-3
[17] Chen W C, Gentile T R, Walker T G and Babcock E 2007 *Phys. Rev.* A **75** 013416
[18] Heil W, Andersen K, Hofmann D, Humblot H, Kulda J, Lelievre-Berna E, Schärf O and Tasset F 1998 *Physica* B*: Condensed Matter* **241-243** 56-63
[19] Fitzsimmons W A, Tankersley L L and Walters G K 1969 *Phys. Rev.* **179** 156-165
[20] Jacob R E, Morgan S W and Saam B 2001 *Phys. Rev .Lett.* **87** 143004
[21] Schmiedeskamp J, Heil W, Otten E W, Kremer R K, Simon A and Zimmer J 2006 *Eur. Phys. J.* D **38** 427-38
[22] Deninger A, Heil W, Otten E W, Wolf M, Kermer R K and A.Simon A 2006 *Eur. Phys. J.* D **38** 439-43
[23] Schmiedeskamp J, Elmers H J, Heil W, Otten E W, Sobolev Yu, Kilian W, Rinneberg H, Sander-Thömmes T, Seifert F and Zimmer J 2006 *Eur.Phys. J.* D **38** 445-54
[24] Newbury N R, Barton A S, Cates G D, Happer W and Middleton H 1993 *Phys. Rev.* A **48** 4411-20
[25] Gamblin R L and Carver T R 1965 *Phys. Rev.* **138** (4A) 946-60
[26] Schearer L D and Walters G K 1965 *Phys. Rev.* **139** (5A) 1398-402
[27] Cates G D, Schaefer S R and Happer W 1988 *Phys. Rev.* A **37** 2877-85
[28] Dreyer J, Regnault L P, Bourgeat-Lami E, Lelievre-Berna E, Pujol S, Thomas F, Thomas M and Tasset F 2000 *Nucl. Instr. Meth.* A **449** 638 -48
[29] Ino T and Muto S 2005 *Nucl. Instr. Meth.* A **548** 499 -506
[30] Petoukhov A K, Guillard V, Andersen K H, Bourgeat-Lami E, Chung R, Humblot H, Jullien D, Lelièvre-Berna E, Soldner T, Tasset F and Thomas M 2006 *Nucl. Instr. Meth.* A **560** 480-4
[31] Hutanu V, Meven M, Sazonov A and Heger G 2008 *Meas. Sci. Technol.* **19** 034010
[32] Mclver J W, Erwin R, Chen W C and Gentile T R 2009 *Rev. Sci. Instr.* **80** 063905
[33] Hiebel S, Großmann T, Kiselev D, Schmiedeskamp J, Gusev Y, Heil W, Karpuk S, Krimmer J, Otten E W and Salhi Z 2010 *Journal of Magnetic Resonance* **204** 37-49
[34] URL http://femm.berlios.de
[35] Bigelow N P, Nacher P J and Leduc M 1992 *J. Phys. II France* **2** 2159-79
[36] Lorenzon W, Gentile T R, Gao H and McKeown R D 1993 *Phys. Rev.* A **47** 468-479

Journal of Physics: Conference Series **294** (2011) 012017 doi:10.1088/1742-6596/294/1/012017

Applications of ^{3}He neutron spin filters on the small-angle neutron scattering spectrometer SANS-J-II

Y Sakaguchi[1,8] *, H Kira[1], T Oku[1], T Shinohara[1], J Suzuki[1,8], K Sakai[1], M Nakamura[1], K Aizawa[1], M Arai[1], Y Noda[1], S Koizumi[1], M Takeda[1], Y Endoh[1], L J Chang[1, 2, 3], Y Arimoto[4], T Ino[4], H M Shimizu[4], T Kamiyama[4], K Ohoyama[5], H Hiraka[5], K Tsutsumi[6], K Yamada[7] and K Kakurai[1]

[1]Japan Atomic Energy Agency, Tokai, Naka-gun, Ibaraki 319-1195, Japan
[2]Department of Physics, National Cheng Kung University, Tainan 70101, Taiwan
[3]Nuclear Science and Technology Development Center, National Tsing Hua University, Hsinchu 30013, Taiwan
[4]Institute of Materials Structure Science, High Energy Accelerator Research Organization (KEK), 1-1 Oho, Tsukuba 305-0801, Japan
[5]Institute for Materials Research, Tohoku University, Aoba-ku, Sendai 980-8577, Japan
[6]Department of Physics, Tohoku University, Aoba-ku, Sendai 980-8578, Japan
[7]WPI Advanced Institute for Materials Research, Tohoku University, Aoba-ku, Sendai 980-8577, Japan

E-mail: y_sakaguchi@cross.or.jp

Abstract. A polarized ^{3}He neutron spin filter has been applied to polarization analysis on the small-angle neutron scattering spectrometer SANS-J-II at JRR-3. Measurements were taken on silver behenate, which has several coherent peaks in the small-angle region with a background of spin incoherent hydrogen scattering. Here we demonstrate that the coherent and spin incoherent scattering were successfully separated by the polarization analysis using the ^{3}He spin filter on the instrument.

1. Introduction

Polarization analysis is a powerful technique to distinguish magnetic scattering from nuclear scattering and to separate spin incoherent scattering from nuclear coherent scattering. In the polarization analysis on a small-angle neutron scattering instrument, ^{3}He neutron spin filters are useful because of their broad angular acceptance, which is limited only by the solid angle subtended by the ^{3}He cell. Also, the ^{3}He cell contributes little scattering in the small-angle region. The ^{3}He neutron spin filters use very high spin-dependent neutron capture cross-sections of ^{3}He gas and high ^{3}He nuclear spin polarization is required for the scattering experiments. Recent progress on ^{3}He polarization techniques enables us to use ^{3}He neutron spin filters in a small-angle scattering instrument [1,2]. In Japan, a high ^{3}He polarization technique based on spin-exchange optical pumping (SEOP) was established by Ino et al [3, 4] and our ^{3}He neutron spin filter development group in Japan is improving on this technique [5, 6]. Current circumstances mean that it is now possible to obtain a supply of polarized ^{3}He gas for the neutron scattering experiment in Japan. In this paper, we report on the applications of ^{3}He neutron spin filters on the neutron small-angle scattering spectrometer

[8] Present address: Research Center for Neutron Science and Technology, Comprehensive Research Organization for Science and Society, Tokai, Ibaraki 319-1106, Japan

Journal of Physics: Conference Series **294** (2011) 012017 doi:10.1088/1742-6596/294/1/012017

SANS-J-II at JRR-3 and demonstrate that ^{3}He neutron spin filters are now available on the instrument. As a test sample, we have chosen silver behenate [7, 8] and tried to separate spin incoherent hydrogen scattering from coherent scattering of the sample.

2. Experiment

2.1 Spectrometer and experimental setup

The experiment was performed with the small-angle neutron scattering spectrometer SANS-J-II at JRR-3. In the spectrometer, polarization devices are fully installed and include an Fe/Si polarising supermirror, a two-coil -flipper, a radio-frequency gradient π-flipper and a supermirror analyzer [9-12]. We used the polarizing supermirror to get a polarized incident neutron beam. The spin of the beam was flipped by a two-coil flipper. The scattered neutron beams were detected by 25 cm x 25 cm high-angle ^{3}He detectors. The arrangement of the experimental apparatus is illustrated in Figure 1(a). We used a 4 cm diameter, sealed spin-exchange cell made of GE180 glass [13]. It was placed in a 54 cm long and 12 cm inner diameter solenoid coil. In order to get the maximum possible coverage of the scattering angle region, the sample was placed 17 cm behind the cell in the solenoid coil, as shown in Fig.1(b). The spin filter cell was filled with 6.7 atm-cm of ^{3}He gas, which was chosen considering the neutron wavelength of the spectrometer, 6.5 Å. The spin-exchange cell was pumped up at our optical pumping station, and then, it was transported to SANS-J-II. To avoid unwanted scattering from the glass cell and the solenoid coil, the inside of the coil and the space around the cell were covered by B_4C rubber.

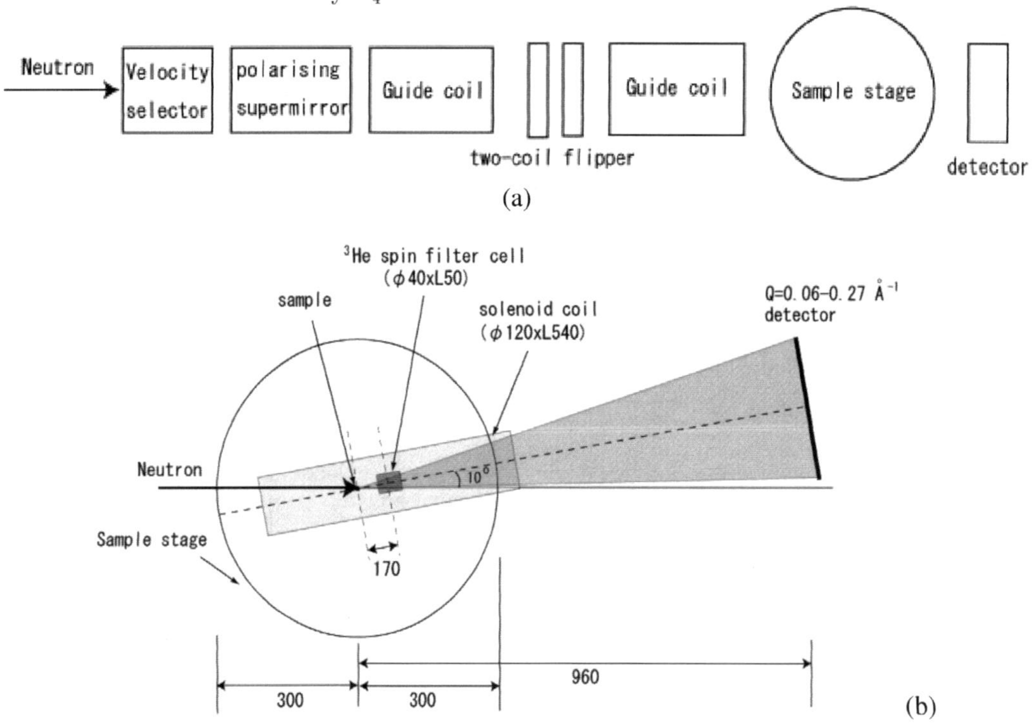

Figure 1. Arrangement of experimental apparatus.

2.2 Sample

In this experiment, we have chosen silver behenate as the test sample. This is a crystalline long-chain silver carboxylate, $CH_3(CH_2)$ COOAg. There is a (001) long-period spacing of 58.38 Å, which results from a tail-to-tail alignment of two silver behenate molecules. Due to the presence of order, the material is used for low-angle calibration of X-ray and neutron diffractometers in a Q-range of order 10^{-1}Å^{-1}. In the previous SANS experiment by Gilles et al., which was performed with an unpolarized neutron beam, an intensive first peak,

Journal of Physics: Conference Series **294** (2011) 012017 doi:10.1088/1742-6596/294/1/012017

comparatively small 2nd and 3rd peaks, and a constant background component were observed. The ratio of these components is 1.00 : 0.12 : 0.06 : 0.29. The aim of this experiment is to subtract this background component, which would originate from spin incoherent hydrogen scattering.

3. Results and discussion

3.1. Neutron small-angle scattering for unpolarized and polarized beams

Figure 2 shows the small-angle neutron scattering profiles of silver behenate, which was obtained from an unpolarized neutron beam. The sample was placed in the solenoid coil, but the ^{3}He neutron spin filter was not. The sample stage was fixed at 10°. The pattern is similar to that of Gilles et al., in which a strong first peak, weak second peak and constant background component are observed. The ratio of the first peak, the second peak and the background component is 1.00 : 0.1 : 0.23. This is consistent with the result measured by Gilles et al.

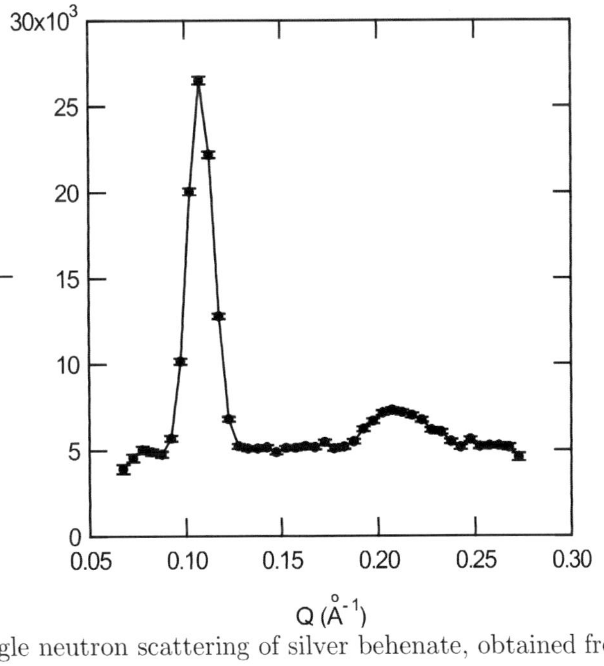

Figure 2. The small-angle neutron scattering of silver behenate, obtained from an unpolarized neutron beam.

Figure 3 shows the result obtained from a polarized neutron beam. To obtain the data, the absorption of the sample, the background measured by a scattering from the empty sample cell, and the sensitivity of the detector were corrected. There was a large loss in intensity compared to the measurement using an unpolarized beam. However, the difference in scattering pattern is clearly observed between "spin flipper on" and "off": there are two peaks and a constant Q-independent component for "spin flipper on", while there is only the constant Q-independent component for "spin flipper off". From the result, it is found that the scattering for "spin flipper on" represents the scattering when the ^{3}He polarization is parallel to incident neutron polarization, while scattering for "spin flipper off" represents scattering when the ^{3}He polarization is antiparallel to the incident neutron polarization.

JCNS Workshop on Modern Trends in Production and Applications of Polarized ^{3}He IOP Publishing
Journal of Physics: Conference Series **294** (2011) 012017 doi:10.1088/1742-6596/294/1/012017

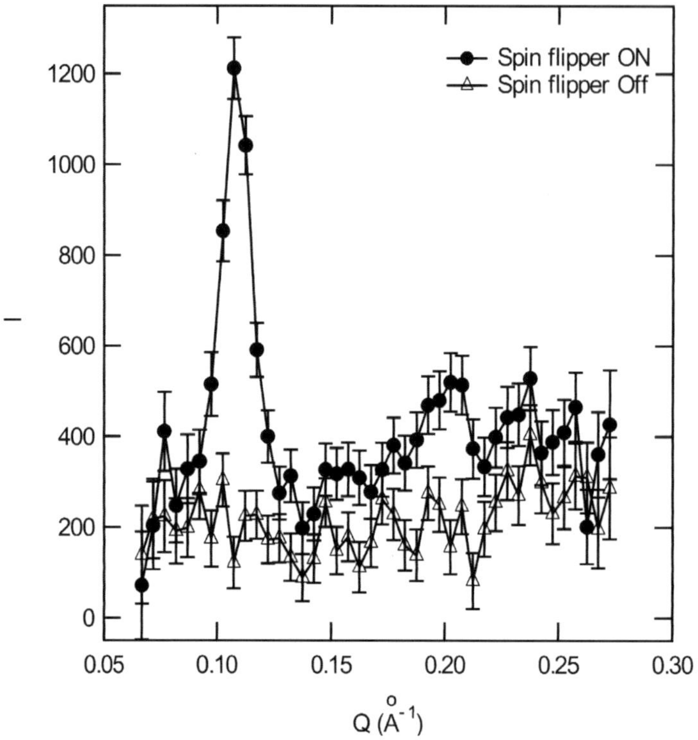

Figure 3. Polarized small-angle neutron scattering of silver behenate.

3.2. ^{3}He polarization

The ^{3}He polarization has been checked by measuring the transmittance through the ^{3}He filter during the experiment. The initial ^{3}He polarization was 34 % and the relaxation time was 57 h. These values were less than those we could expect from previous measurements [14]. Because of the fast decay in ^{3}He polarization, we expect that there was a disturbance in the magnetic environment during the experiment.

3.3. Separation of spin incoherent scattering from nuclear coherent scattering

The imperfection of spin-filtering by the ^{3}He spin filter has to be corrected considering the incident neutron polarization and the transmission of neutrons through the ^{3}He spin filter [15, 16]. For the incident neutron beam, the polarization efficiency of the supermirror polarizer is better than 0.99 and the flipping efficiency of the two-coil flipper is almost unity [12]. Therefore, we assume here that the polarization of the incident neutron beam is unity for both "spin flipper on" and "off". In this case, the observed number of neutrons scattered by the sample for the incident neutron beam polarization parallel (anti-parallel) to the ^{3}He polarization, N_+ (N_-), is expressed as:

$$N_+ = N_{\mathrm{nsf}}T_+ + N_{\mathrm{sf}}T_- \quad \text{and}$$

$$N_- = N_{\mathrm{nsf}}T_- + N_{\mathrm{sf}}T_+ \tag{1}$$

where T_+(T_-) is the transmission of the ^{3}He analyzer for neutron spin parallel (antiparallel) to the ^{3}He polarization. T_+ and T_- are given by

$$T_+ = \exp\{-\sigma_0(1-P_{\mathrm{He}})\rho l\} \quad \text{and}$$

$$T_- = \exp\{-\sigma_0(1+P_{\mathrm{He}})\rho l\} \tag{2}$$

where σ_0 is the absorption cross-section for unpolarized neutrons, ρ is the number density of the ^{3}He gas, and l is the thickness of the ^{3}He neutron spin filter. For ^{3}He polarization, P_{He}, we used the value obtained in section 3.2.
From the above equations, we obtain:

$$N_{nsf} = (1/4)\exp(-\sigma_0\rho l)\{(N_+ + N_-)/\cosh(\sigma_0 P_{He}\rho l) + (N_+ - N_-)/\sinh(\sigma_0 P_{He}\rho l)\}$$

$$N_{sf} = (1/4)\exp(-\sigma_0\rho l)\{(N_+ + N_-)/\cosh(\sigma_0 P_{He}\rho l) - (N_+ - N_-)/\sinh(\sigma_0 P_{He}\rho l)\} \quad (3)$$

Using the relationship between non-spin-flip scattering, spin-flip scattering, coherent scattering and spin-incoherent scattering [17, 18]:

$$N_{nsf} = (1/3)N_{si} + N_c$$

$$N_{sf} = (2/3)N_{si} \quad (4)$$

where the subscripts refer to coherent (c) and spin-incoherent (si) scattering. N_c and N_{si} are finally written as:

$$N_c = (1/8)\exp(\sigma_0\rho l)\{(N_+ + N_-)/\cosh(\sigma_0 P_{He}\rho l) + 3(N_+ - N_-)/\sinh(\sigma_0 P_{He}\rho l)\}$$

$$N_{si} = (3/8)\exp(\sigma_0\rho l)\{(N_+ + N_-)/\cosh(\sigma_0 P_{He}\rho l) - (N_+ - N_-)/\sinh(\sigma_0 P_{He}\rho l)\} \quad (5)$$

Figure 4 is the result of separation of the coherent and incoherent scattering components, in which these corrections are made. Although the data are scattered, the ratio of the intensity between the first peak and the incoherent component is about 1.00 : 0.23. This is consistent with the value obtained from an unpolarized neutron beam.

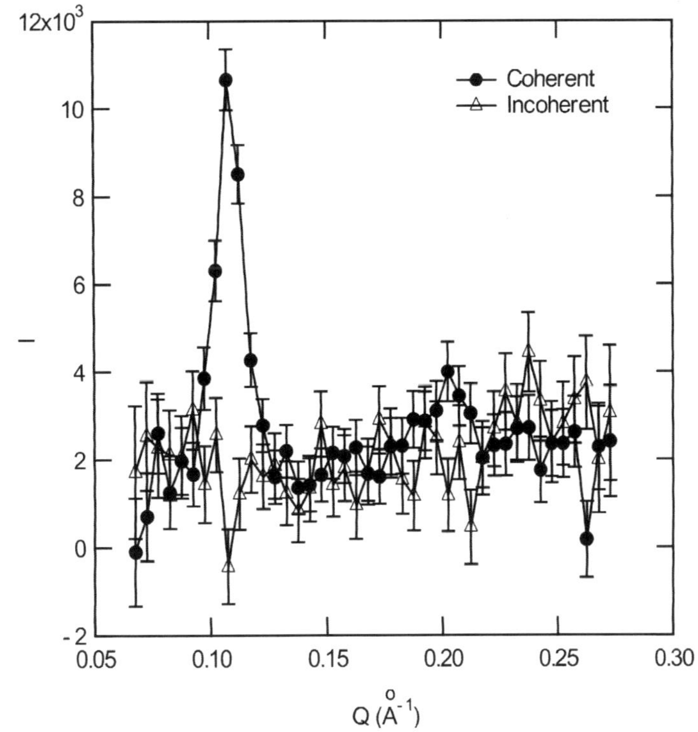

Figure 4. Separation of coherent and incoherent scattering for silver behenate.

4. Summary

We have applied ^{3}He spin filters to polarization analysis on the small-angle neutron scattering spectrometer SANS-J-II. For the silver behenate test sample, the spin incoherent hydrogen scattering was separated from coherent scattering. From the results, it was shown that ^{3}He spin filters can contribute to the upgrade in the focusing and polarized small-angle neutron scattering spectrometer, SANS-J-II. This method is applicable to other protonated polymers or biological samples where the presence of massive spin-incoherent scattering from hydrogen can hinder the determination of the coherent scattering. Also, we expect that this demonstrative experiment encourages further polarization analysis with other neutron scattering instruments at JRR-3 and J-PARC (Japan Proton Accelerator Research Complex). In the present experiment, the decay of ^{3}He polarization was faster than expected. This could be caused by the conditions in the magnetic environment being insufficient and should be improved for our subsequent studies.

Acknowledgments

This research project has been supported by the Quantum Beam Fundamental Development program, MEXT. The authors acknowledge T. Kudo of the glass shop in Institute of Multidisciplinary Research for Advanced Materials, Tohoku University, for construction of the cell.

References

[1] For example: Chen W C, Erwin R, Mclver III J W, Watson S, Fu C B, Gentile T R, Borchers J A, Lynn J W and Jones G L 2009 *Physica B: Condensed matter* **404** 2663
[2] Andersen K H, Jullien D, Petoukhov A K, Mouveau P, Bordennave F, Thomas F and Babcock E 2009 *Physica B: Condensed matter* **404** 2652
[3] Ino T, Muto S, Masuda Y, Kim G and Skoy V R 2005 *Physica B: Condensed matter* **356** 109
[4] Ino T, Nakamura M, Oku T, Shinohara T, Suzuki J, Ohoyama K and Hiraka H 2009 *Physica B: Condensed matter* **404** 2667
[5] Kira H et al., *IOP Journal of Physics Conference Series, this volume* and *Physica B: Condensed matter* , in press, doi :10.1016/j.physb.2010.12.019
[6] Arimoto Y et al., *Physica B: Condensed matter*, in press, doi :10.1016/j.physb.2010.11.089
[7] Blanton T N, Barnes C L and M. Lelental 2000 *J. Appl. Cryst.* **33** 172
[8] Gilles R, Keiderling U and Wiedenmann 1998 *J. Appl. Crystallog.* **31** 957
[9] Koizumi S, Iwase H, Suzuki J, Oku T, Motokawa R, Sasao H, Tanaka H, Yamaguchi D, Shimizu H M and Hashimoto T 2006 *Physica B: Condensed matter* **385-386** 1000
[10] Iwase H, Koizumi S, Suzuki J, Oku T, Sasao H, Tanaka H, Shimizu H M and Hashimoto T 2007 *J. Appl. Crystallog.* **40** s414
[11] Koizumi S, Iwase H, Suzuki J, Oku T, Motokawa R, Sasao H, Tanaka H, Yamaguchi D, Shimizu H M and Hashimoto T 2007 *J. Appl. Crystallog.* **40** s474
[12] Oku T, Iwase H, Shinohara T, Yamada S, Hirota K, Koizumi S, Suzuki J, Hashimoto T and Shimizu H M 2007 *J. Appl. Crystallog.* **40** s408
[13] GE Lighting Component Sales, Bldg. 315D, 1975 Noble Rd., Cleveland, OH 44117
[14] When some of the authors (Ohoyama K, Tsutsumi K and Ino T) conducted polarized neutron scattering experiment using a ^{3}He spin filter on HERMES at JRR-3, the initial ^{3}He polarization was 60 % and the relaxation time was 200 h
[15] Gentile T R, Jones G L, Thompson A K, Barker J, Glinka C J, Hammouda B and Lynn J W 2000 *J. Appl. Crystallog.* **33** 771
[16] Krycka K L, Booth R, Borchers J A, Chen W C, Conlon C, Gentile T R, Hogg C, Ijiri Y, Laver M, Maranville B B, Majetich S A, Rhyne J J and Watson S M 2009 *Physica B: Condensed matter* **404** 2561

[17] Moon R M, Riste T and Koehler 1969 *Phys. Rev.* **181** 920
[18] Dore J C, Clarke J H and Wenzel J T 1976 *Nucl. Instrum. Meth.* **138** 317

AUTHOR INDEX